THE ESSENTIAL WELDER: GAS METAL ARC WELDING CLASSROOM MANUAL

THE ESSENTIAL WELDER: GAS METAL ARC WELDING CLASSROOM MANUAL

Richard Rowe
Larry Jeffus

Africa • Australia • Canada • Denmark • Japan • Mexico • New Zealand • Philippines •
Puerto Rico • Singapore • Spain • United Kingdom • United States

NOTICE TO THE READER

Delmar Staff

Business Unit Director: Alar Elken
Executive Editor: Sandy Clark
Acquisitions Editor: Vernon R. Anthony
Editorial Assistant: Bridget Morrison
Executive Marketing Manager: Maura Theriault

Channel Manager: Mona Caron
Marketing Coordinator: Kasey Young
Executive Production Manager: Mary Ellen Black
Project Editor: Barbara L. Diaz

COPYRIGHT © 2000
Delmar is a division of Thomson Learning. The Thomson Learning logo is a registered trademark used herein under license.

Printed in the United States of America
1 2 3 4 5 6 7 8 9 10 XXX 05 04 03 02 01 00

For more information, contact Delmar at 3 Columbia Circle, PO Box 15015, Albany NY 12212-5015; or find us on the World Wide Web at
http://www.delmar.com

Asia
Thomson Learning
60 Albert Street, #15-01
Albert Complex
Singapore 189969

Australia/New Zealand
Nelson/Thomson Learning
102 Dodds Street
South Melbourne, Victoria 3205
Australia

Canada
Nelson/Thomson Learning
1120 Birchmount Road
Scarborough, Ontario
Canada M1K 5G4

International Headquarters
Thomson Learning
International Division
290 Harbor Drive, 2nd Floor
Stamford, CT 06902-7477
USA

Japan
Thomson Learning
Palaceside Building 5F
1-1-1 Hitotsubashi, Chiyoda-ku
Tokyo 100 0003
Japan

Latin America
Thomson Learning
Seneca, 53
Colonia Polanco
11560 Mexico D.F. Mexico

Spain
Thomson Learning
Calle Magallanes, 25
28015-Madrid
Espana

UK/Europe/Middle East
Thomson Learning
Berkshire House
168-173 High Holborn
London
WC1V 7AA United Kingdom

Thomas Nelson & Sons Ltd.
Nelson House
Mayfield Road
Walton-on-Thames
KT 12 5PL United Kingdom

Library of Congress Cataloging-in-Publication Data

Rowe, Richard J. (Richard James), 1947–
    The essential welder : gas metal arc welding : classroom manual / Richard J. Rowe.
      p.   cm.
    Includes index.
    ISBN: 0–8273–7608–1
    1. Gas metal arc welding—Handbooks, manuals, etc. I. Title.
TK4660.R68 1999
671.5'22—dc21

98–29584
CIP

CONTENTS

PREFACE

Gas metal arc welding (GMAW) is one of the easiest welding processes to learn once the equipment is set up. Often new welders can make some of the basic welds within the first hour and even master several by the end of the day. The key to making good welds is to properly set up the welder. Once you master the settings and can control the weld pool, you have most of the skills required of a GMA production welder.

In this textbook we focus on helping you master the essentials of this process as quickly as possible. Once you have attained these skills, further developments in your welding abilities are practically unlimited. GMA welding is one of the most popular industrial welding processes. Its popularity comes from both its high productivity and the high quality of the welds it can produce.

Because many of the skills you will develop while making flat-position welds on mild steel are easily transferred to other positions and metals, we will start with these welds. After you have learned the basics, you will move on to other positions and metals such as stainless steel and aluminum.

The lab manual that accompanies this classroom manual provides in-depth exercises as well as useful and practical projects using welding skills. Other exercises hone your math, reading, writing, and English skills. In this way the lab manual gets you involved and makes learning welding fun!

Acknowledgments

I would like to express my sincere thanks and appreciation to some of the people who helped make this book possible: Carol Jeffus for proofreading, Tina Ivey for typing and organizing of material, and Wendy Jeffus, Amy Jeffus, Jason Madore, and Ted Robinson for their work on the art manuscript.

Dedication

This book is dedicated to three very special people—my wife Carol and my daughters Wendy and Amy.

CHAPTER 1

INTRODUCTION TO GMA WELDING

A *weld* is defined by the American Welding Society (AWS) as "a localized coalescence (the fusion or growing together of the grain structure of the materials being welded) of metals or nonmetals produced either by heating the materials to the required welding temperatures, with or without the application of pressure, or by the application of pressure alone, and with or without the use of filler materials." *Welding* is defined as "a joining process that produces coalescence of materials by heating them to the welding temperature, with or without the application of pressure or by the application of pressure alone, and with or without the use of filler metal. In less technical language, a weld is made when separate pieces of material combine and form one piece when heated to a temperature high enough to cause softening or melting and flow together. Pressure may or may not be used to force the pieces together. In some instances, pressure alone may be sufficient to force the separate pieces of material to combine and form one piece. Filler material is added when needed to form a completed weld in the joint. It is also important to note that the word *material* is used because today welds can be made from a growing list of materials such as plastic, glass, and ceramics.

GMA WELDING

GMA welding uses a welding wire that is fed automatically at a constant speed as an electrode. An arc is generated between the base metal and the wire, and the resulting heat from the arc melts the welding wire and joins the base metals, (Figure 1-1).

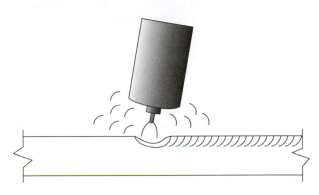

Figure 1-1

This method is known as a semiautomatic arc-welding process because wire is fed automatically at a constant rate and the welder provides gun movement. During the welding process, a shielding gas protects the weld from the atmosphere and prevents oxidation of the base metal. The type of shielding gas used depends on the base material to be welded.

This process derives its name from the fact that originally the process used only inert gases for shielding, so the name metal inert gas (MIG) welding applied. Today many different gases are used. Some are inert and nonreactive under all conditions, and others are reactive and can combine under some conditions. Because of the changes in the gas shielding, the term *gas metal arc-welding (GMAW)* was adopted by the American Welding Society for this

1

process. However, in the auto body industry the term *MIG* is more commonly used. This process has had other names through the years, such as wire welding, but regardless of the name, the process is the same.

The advantages of GMA welding over conventional electrode-type arc-(stick) welding are numerous. Car manufacturers, insurance companies, I-CAR, and governmental regulations require or recommend that GMA welding be used in virtually all welding repairs. The advantages of GMA welding are as follows:

- GMA welding is easy to learn. The typical welder can learn to use GMA welding equipment in just a few hours of instruction and practice. More time may be required to master the adjusting of the equipment.

- GMA welding can produce higher-quality welds faster and more consistently than conventional stick electrode welds.

- Low current can be used to GMA weld thin metals.

- Fast welding speeds and low currents prevent heat damage to adjacent areas that can cause strength loss and warping.

- The small molten weld pool is easily controlled.

- GMA welding is tolerant of gaps and misfits. Gaps can be welded by making several spot welds on top of each other.

- Almost all auto body steels can be GMA welded with one common type of weld wire.

- Metals of different thicknesses can be GMA welded with the same diameter of wire.

- With GMA welding, vertical and/or overhead welding is possible because the weld pool is small and the metal is molten for only a very short time.

- GMA welds are easily started in the correct spot because the wire is not energized until the gun trigger is depressed.

- GMA welding produces minimum waste of welding consumables.

GMA welding became popular when auto manufacturers began using high-strength steel (HSS) and high-strength low-alloy (HSLA) steel. These materials are strong enough to be used in much thinner gauges than had been used in the past. The only correct way to weld HSS, HSLA, and other thin-gauge steel is with GMA welding. Welding a rear-quarter panel with an oxyacetylene welder takes about 4 hours; a GMA welder can do the same job in about 30 minutes.

GMA welding is not limited to body repairs alone. It is also ideal for exhaust-pipe welding, repairing mechanical supports, and installing trailer hitches and bumpers. Almost any welding that would be done with either an arc or a gas welder can be done faster with GMA welding. In addition, it is possible to weld aluminum sheet and aluminum castings such as cracked transmission cases, cylinder heads, and intake manifolds.

EQUIPMENT

The basic GMA welding equipment consists of the gun, electrode (wire) feed unit, electrode (wire) supply, power source, shielding gas supply with flowmeter/regulator, control circuit, and related hoses, liners, and cables (Figure 1-2). The system should be portable (Figure 1-3). In some cases, the system can be used for more than one process. These power sources can be switched over for other uses.

Power Source

The power source consists of a transformer and a rectifier. They produce a DC welding current ranging from 40–600 amperes with 10–40 volts, depending upon the machine. In the past, some GMA welding processes used Alternating Current (AC) welding current, but Direct Current Electrode Positive (DCEP) is used exclusively for all GMA work today. Typical power supplies are shown in Figure 1-4.

Because of the long periods of continuous use, GMA welding machines have a 100% duty cycle, which allows the machine to be run continuously without damage.

Wire Feed Unit

The wire feeder provides a steady and reliable supply of wire to the weld (Figure 1-5). Slight changes in the rate at which the wire is fed have distinct effects on the weld.

The motor used in a feed unit can be continuously adjusted over the desired range.

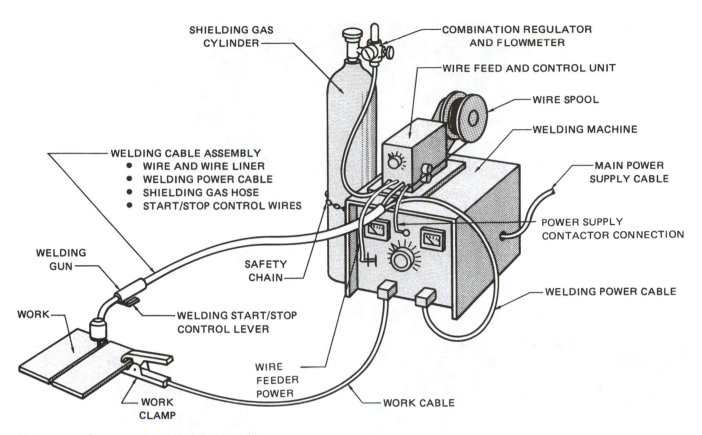

Figure 1-2 Gas metal arc-welding equipment.

Figure 1-3 Portable GMA welding. (Courtesy of Miller Electric Mfg. Co.)

(A)

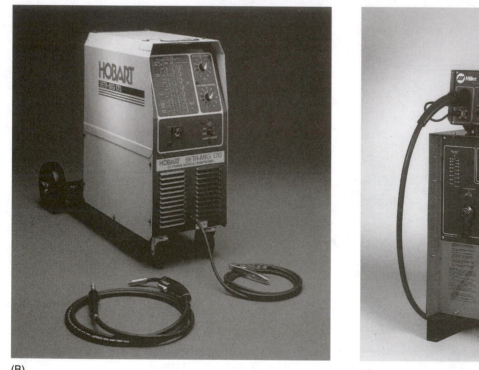

(B)

(C)

Figure 1-4 (A) Courtesy of Lincoln Electric, Cleveland, OH. (B), (C) Courtesy of Miller Electric Mfg. Co.

Figure 1-5 Large-capacity wire feed unit used with GMAW. (Courtesy of the Lincoln Electric Company)

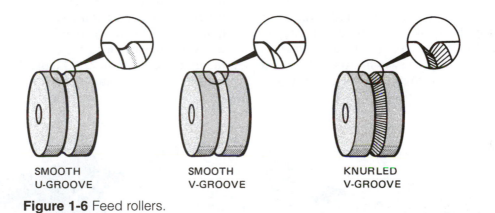

SMOOTH
U-GROOVE

SMOOTH
V-GROOVE

KNURLED
V-GROOVE

Figure 1-6 Feed rollers.

Push-Type Feed System

The wire rollers are clamped securely against the wire to provide the necessary friction to push the wire through the conduit to the gun. The pressure applied on the wire can be adjusted. A groove is provided in the roller to aid in alignment and to lessen the chance of slippage. Most manufacturers provide rollers with smooth or knurled U-shaped or V-shaped grooves (Figure 1-6). Knurling (a series of ridges cut into the groove) helps grip larger-diameter wires so that they can be pushed along more easily. Soft wires, such as aluminum, are easy to damage if knurled rollers are used. Aluminum wires are best used with U-grooved rollers. Even V-grooved rollers can distort the surface of soft wire, causing problems. V-grooved rollers are best suited for hard wires, such as mild steel and stainless steel. It is also important to use the correct size grooves in the rollers.

In the push-type system, the electrode must have enough strength to be pushed through the conduit without kinking. Mild steel and stainless steel can be readily pushed 15–20 feet (4–6 m), but aluminum is much harder to push over 10 feet (3 m).

Pull-Type Feed System

In pull-type systems, a smaller but higher-speed motor is located in the gun to pull the wire through the conduit. Using this system, it is possible to move even soft wire over great distances. The disadvantages are that the gun is heavier and more difficult to use, rethreading the wire takes more time, and the operating life of the motor is shorter.

Spool Gun

A spool gun is a compact, self-contained system consisting of a small drive system and a wire supply (Figure 1-7). This system allows the welder to move freely around a job with only a power lead and shielding gas hose to manage. The major control system is usually mounted on the welder. The feed rollers and motor are found in the gun just behind the nozzle and contact tube. Because of the short distance the wire must be moved, very soft wires (aluminum) can be used. A small spool of welding wire is located just behind the feed rollers. The small spools of wire required in these guns are often very expensive. Although the guns are small, they feel heavy when being used.

Electrode Conduit

The electrode conduit or liner guides the welding wire from the feed rollers to the gun. It may be encased in a lead that contains the shielding gas.

Power cable and gun switch circuit wires are contained in a conduit that is made of a tightly wound coil having the needed flexibility and strength. The steel conduit may have a nylon or Teflon® liner to protect soft, easily scratched metals, such as aluminum, as they are fed.

If the conduit is not an integral part of the lead, it must be firmly attached to both ends of the lead. Failure to attach the conduit can result in misalignment, which causes additional drag or makes the wire jam completely. If the conduit does not extend through the lead casing to make a connection, it can be drawn out by tightly coiling the lead (Figure 1-8). Coiling will force the conduit out so that it can be connected. If the conduit is too long for the lead, it should be cut off and filed smooth. Too long a lead will bend and twist inside the conduit, possibly causing feed problems.

Welding Gun

The welding gun attaches to the end of the power cable, electrode conduit, and shielding gas hose. It is used by the welder to produce the weld. A trigger switch starts and stops the weld cycle. The gun also has a contact tube Figure 1-9(2b) that transfers welding current to the electrode moving through the gun and a gas nozzle Figure 1-9(1b) that directs the shielding gas onto the weld.

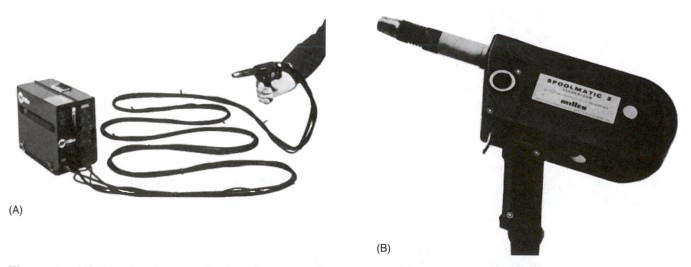

(A)

(B)

Figure 1-7 (A) Wire-feed system that enables the wire to be moved through a longer cable. (Courtesy of Miller Electric Manufacturing Company) (B) Feeder/gun for GMA welding. (Courtesy of Miller Electric Mfg. Co.)

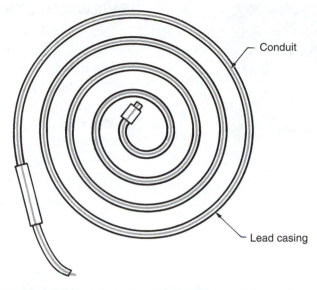

Figure 1-8 Tightly coiled lead casing will force the liner out of the gun.

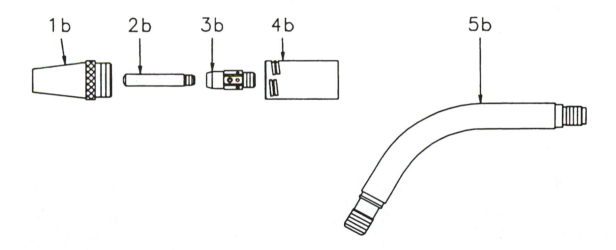

Figure 1-9 Gun parts. (Courtesy of Tweco Products Inc.)

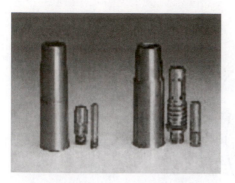

Figure 1-10

REVIEW QUESTIONS

1. Define *welding*.
2. When is filler material added?
3. What is the purpose of a shielding gas?
4. What does MIG stand for?
5. When did GMA welding become popular?
6. What does the basic GMA welding equipment consist of?
7. What is the purpose of the wire feeder?
8. What is a welding gun?

CHAPTER 2

SAFETY

All welding processes pose potential safety problems. When performed in conjunction with a motor vehicle, welding can present additional hazards and concerns. To avoid injuries to individuals and damage to property during welding, welders must follow strict safety measures and precautions.

All federal, state, and local laws, codes, standards, and regulations related to the welding industry must be followed. These laws, rules, and regulations address a variety of issues related to welding, such as health, the environment and personal and public safety. Many health issues are concerned with the fumes created by welding. Safety topics address concerns such as potential fire hazards and electrical hazards as well as the safe operation of repaired vehicles. Environmental issues deal with air quality, water quality, and hazardous waste generated by welding. The welding shop operator is responsible for making certain that all such concerns are addressed appropriately. In addition to these governmental regulations, there are also many requirements established by insurance companies that provide insurance coverage for your business. As part of their coverage, many insurance companies will assist a welding shop in establishing appropriate operating procedures that will ensure that all work performed meets current safety standards.

Many trade and professional associations offer recommendations for the various welding processes. The American Welding Society (AWS), American National Standards Institute (ANSI), American Society for Testing and Materials (ASTM), National Fire Protection Association (NFPA), Compressed Gas Association (CGA), and National Safety Council (NSC) have established many voluntary safety standards for the welding trade. Although these organizations' standards are not laws, they are used by most governmental regulatory agencies in setting their legal requirements.

Although there are serious threats to safety in the welding shop, welding can be a safe occupation if proper precautionary measures are instituted and scrupulously followed. Under these situations, welding is no more hazardous or injurious to health than any other metal-working occupation.

BURNS

Burns are one of the most common and painful injuries that occur in the auto body shop. Burns can be caused by ultraviolet light rays as well as by contact with hot material. The chance of infection from a burn is high because of the dead tissue. To reduce the chance of infection, all burns must receive proper medical treatment. Burns are divided into three classifications, depending upon the degree of severity. These categories are first-degree, second-degree, and third-degree burns.

First-Degree Burns

First-degree burns occur when the surface of the skin is reddish in color, tender, and painful and the skin is unbroken. The first step in treating a first-degree burn is to immediately put the burned area under cold (not icy) water or apply cold-water compresses (clean towel, washcloth, or handkerchief soaked in cold water) until the pain decreases. Then cover the area with sterile bandages or a clean cloth. Do not apply butter or grease or any other home remedies or medications without a doctor's recommendation (Figure 2-1).

Second-Degree Burns

Second-degree burns have occurred when the surface of the skin is severely damaged, resulting in the formation of blisters and possible breaks in the

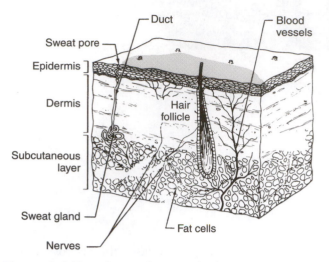

Figure 2-1 First-degree burn: only the skin surface (epidermis) is affected.

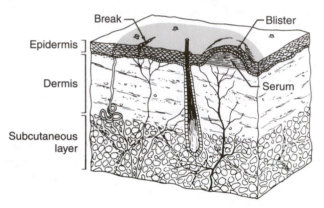

Figure 2-2 Second-degree burn: the epidermis layer is damaged, forming blisters or shallow breaks.

skin. Again, the most important first step is to put the area under cold (not icy) water or apply cold-water compresses until the pain decreases. Gently pat the area dry with a clean towel, and cover the area with a sterile bandage or clean cloth to prevent infection. Seek medical attention. If the burns are around the mouth or nose or involve singed nasal hair, breathing problems may develop. Do not apply ointments, sprays, antiseptics, or home remedies. Reduce the skin temperature as quickly as possible to reduce tissue damage (Figure 2-2).

Third-Degree Burns

Third-degree burns have occurred when the surface of the skin and possibly the tissue below the skin appear white or charred. Initially little pain may be present because nerve endings have been destroyed.

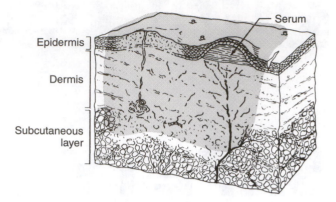

Figure 2-3 Third-degree burn: the epidermis and the subcutaneous layers of tissue are destroyed.

Do not remove any clothes that are stuck to the burn. Do not put ice water or ice on the burns because this could intensify the shock reaction. Do not apply ointments, sprays, antiseptics, or home remedies to these burns. If the victim is on fire, smother the flames with a blanket, rug, or jacket. Breathing difficulties are common with burns around the face, neck, and mouth. Be sure that the victim is breathing. Place a cold cloth or cool (not icy) water on burns of the face, hands, or feet. Cover the burned area with thick, sterile, nonfluffy dressings. Call for an ambulance immediately. People with even small third-degree burns need to consult a doctor (Figure 2-3).

Burns Caused by Light

Some types of light (including ultraviolet, infrared, and visible) can cause burns. Ultraviolet and infrared are not visible to the unaided human eye. During welding, one or more of the three types of light may be present. Arc welding produces all three, but gas welding produces only visible and infrared light.

Light from the welding process can be reflected from walls, ceilings, floors, or other large surfaces and is as dangerous as direct welding light. To reduce the danger from reflected light, the welding area, if possible, should be painted flat black, which will reduce the reflected light by absorbing more of it than any other color. When the welding is to be done on a job site or in a large shop or other area that cannot be painted, weld curtains can be placed to absorb the welding light (Figure 2-4). These special portable welding curtains may be either transparent or opaque. Transparent welding curtains are made of a special high-temperature, flame-resistant plastic that will prevent harmful light from passing through.

Ultraviolet Light

Ultraviolet light waves are the most dangerous because they can cause first-degree and second-degree burns to a welder's eyes or exposed skin. Because a welder cannot see or feel ultraviolet light, he or she must stay protected when standing near arc-welding processes. The closer a welder is to the arc and the higher the current, the quicker a burn may occur. The ultraviolet light is so intense during some welding processes that a welder's eyes can receive a flash burn within seconds, and the skin can be burned within minutes. Ultraviolet light can pass through loosely woven, thin, or light-colored clothing and damaged or poorly maintained arc-welding helmets.

Infrared Light

Infrared light is the light wave that is felt as heat. Although infrared light can cause burns, a person will immediately feel this type of light. Therefore, burns can easily be avoided.

Figure 2-4 Portable welding curtains. (Courtesy of Frommelt Safety Products)

Visible Light

Visible light is the light that we see. It is produced in varying quantities and colors during welding. Too much visible light may cause temporary night blindness (poor eyesight under low-light conditions). Too little visible light may cause eye strain, but visible light is not hazardous.

Whether burns are caused by ultraviolet light or hot material, they can be avoided if proper clothing and other protective gear are worn.

FIRE PROTECTION

Fire is a constant danger during welding or cutting. Fire potential cannot always be removed, but it should be minimized. Highly combustible materials should be 35 feet (10.7 m) or more away from any welding. When it is necessary to weld within 35 feet (10.7 m) of combustible materials, when sparks can reach materials farther than 35 feet (10.7 m) away, or when anything more than a minor fire might start, a fire watch is needed.

Fire Watch

A fire watch can be provided by anyone who knows how to sound an alarm and use a fire extinguisher, which must be of the type required to put out a fire of the combustible materials found in the welding area. Combustible materials that cannot be removed from the welding area should be covered by a noncombustible insulating blanket.

Fire Extinguishers

Fire extinguishers (devices capable of putting out fires) must be present during welding, and everyone must know where they are located.

GMAW produces spatter and sparks, and both can travel a considerable distance from the weld zone and be hot enough to start a fire. Check the area for combustibles (things that will burn) both before you start welding and again when the welding task is completed. Contact your local fire department for more information about the requirements of a safe welding environment.

There are four main types of fire extinguishers. Figure 2-5 shows Type A, which is used on combustible materials, such as cardboard, cloth, paper, or wood. Type A fire extinguishers are designated by a green triangle with the letter *A* in the center.

Figure 2-6 illustrates Type B fire extinguishers, which are used on combustible liquids, such as gas, oils, paints, paint thinners, and solvents. Type B fire extinguishers are designated by a red square with the letter *B* in the center.

Figure 2-7 illustrates Type C fire extinguishers, which are used on electrical equipment or electrical fires. The electrical equipment may include motors, welding power sources, and areas where electrical safety devices, known as fuse boxes, are located. Type C fire extinguishers are designated by a blue circle with the letter *C* in the center.

Figure 2-5 Type A fire extinguisher symbol.

Figure 2-6 Type B fire extinguisher symbol.

Figure 2-7 Type C fire extinguisher symbol.

Figure 2-8 shows Type D fire extinguishers, which are used on combustible metals such as magnesium, titanium, or zinc. Type D fire extinguishers are designated by a yellow star with the letter *D* in the center.

CAUTION

Read and understand the instructions on fire extinguishers *before* you strike a welding arc.

Use

A fire extinguisher works by breaking the fire "triangle" of heat, fuel, and oxygen. Most extinguishers both cool the fire and remove the oxygen. Fire extinguishers use a variety of materials to extinguish fires. The major ones found in welding shops use foam, carbon dioxide, a soda-acid gas cartridge, a pump tank, or dry chemicals.

When using a *foam* extinguisher, do not spray the stream directly into the burning liquid. Allow the foam to fall lightly on the fire.

When using a *carbon dioxide* extinguisher, direct the discharge as close to the fire as possible—first at the edge of the flames and gradually at the center.

When using a *dry chemical* extinguisher, direct the extinguisher at the base of the flames. For Type A fires, follow up by directing the dry chemicals at the remaining material that is burning.

Location of Fire Extinguishers

Fire extinguishers should be of the appropriate type for the kinds of combustible materials located nearby (Figure 2-9). The extinguishers should be placed so that they can be easily removed without reaching over combustible material. They should also be placed at a low enough level to be easily lifted off the mounting (Figure 2-10). The location of fire extinguishers should be marked with red paint and signs high enough so that they can be seen from a distance over people and equipment. The extinguishers should also be marked near the floor so that they can be found even if a room is full of smoke (Figure 2-11).

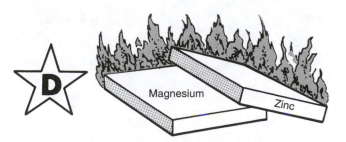

Figure 2-8 Type D fire extinguisher symbol.

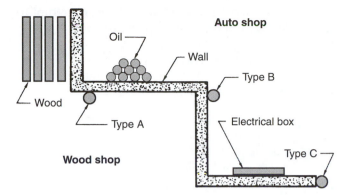

Figure 2-9 The fire extinguisher used should be of the kind that will extinguish the type of flammable material in the area.

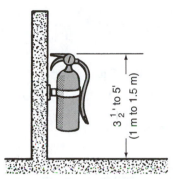

Figure 2-10 The fire extinguisher should be mounted at the correct height so it can be easily lifted.

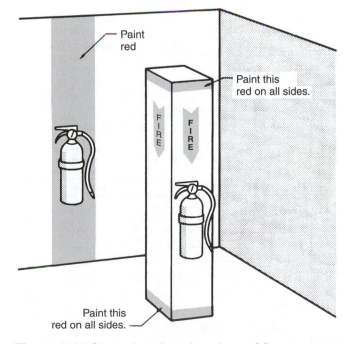

Figure 2-11 Signs that show locations of fire extinguishers should be as large and easily seen as possible.

ELECTRICAL SAFETY

Electrical hazards pose the most life-threatening injuries, but fortunately, they are the least common. By implementing improvements in equipment and establishing safer working conditions, the welding industry has reduced the possibility of electrocution. Not all electrical hazards, however, can be removed, so it is your responsibility to stay alert and follow all safety rules.

To prevent all types of welding shock, you must practice safe welding techniques. Generally there are two areas to consider when dealing with electrical shock: primary voltage and secondary voltage.

Primary voltage is the most dangerous and considered very hazardous because it produces 230–460 volts. Secondary voltage is also dangerous but produces a much lower voltage in the range of 60–100 volts.

ELECTRICAL SAFETY ALERT

Because the primary voltage is so high, only qualified personnel should remove the metal protective cover of the power source for internal maintenance.

CAUTION

Post emergency phone numbers so you can take quick action if problems occur.

The primary voltage power is not turned off when the welding power source on/off switch is in the off position. The power cord, or plug leading from the power supply to the electrical outlet on the wall, must be removed before the primary power is actually off.

The welding equipment has an internal ground (an electrical connection inside the power supply) installed by the qualified technician who set up the machine for welding. This ground connection protects you and the power source if a problem develops inside the equipment. Do not confuse the *internal* ground with the workpiece connection commonly called the *ground clamp*.

If an internal malfunction occurs, a fuse will overheat, melt, and shut off the power, letting you know a problem exists. Never ignore the fact that a machine will blow a fuse.

If the equipment consistently blows fuses after the correct amperage fuse has been installed or replaced, you have a problem. The equipment should be disconnected, properly tagged with an equipment repair tag describing the malfunction or problem, and serviced (Figure 2-12).

Secondary voltage can shock you as well and occurs when you come in contact with the electrical circuit. This means you must touch both the electrode and the work at the same time when you are not insulated and the gun is energized.

CAUTION

Electric shock can kill! Electric shock is the most serious danger you will face when welding. Make it a personal rule *never* to touch or allow others to contact electrically "live" parts.

Always insulate yourself from the work (base metal). Never operate equipment with the protective covers removed. Always disconnect the input power before servicing the equipment. And never attempt to repair electrical equipment if you are not properly trained.

Take a few minutes to inspect the equipment. Locate all the warning labels on the welding machine, and know its potential dangers (Figure 2-13).

CAUTION

Only a qualified repair technician (someone who has been trained to service welding power supplies) should repair welding equipment.

CAUTION

Always keep your body insulated from the electrode and the base metal by using high-quality personal safety equipment.

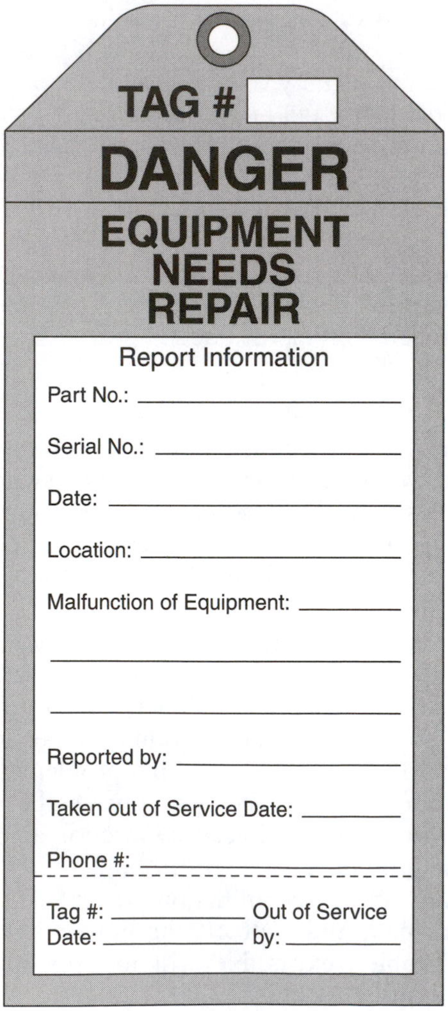

Figure 2-12

Personal Protective Equipment

You are responsible for your own safety when welding. Always use personal protective equipment correctly and consistently.

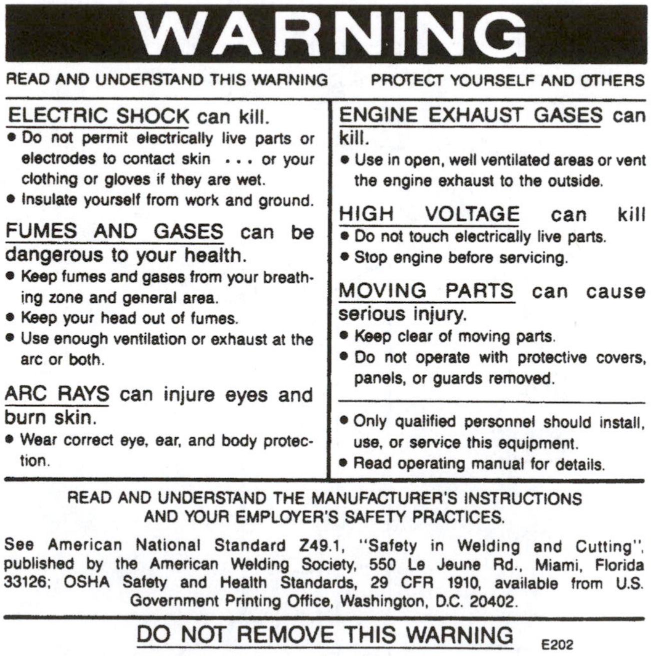

WARNING

READ AND UNDERSTAND THIS WARNING — PROTECT YOURSELF AND OTHERS

ELECTRIC SHOCK can kill.
- Do not permit electrically live parts or electrodes to contact skin . . . or your clothing or gloves if they are wet.
- Insulate yourself from work and ground.

FUMES AND GASES can be dangerous to your health.
- Keep fumes and gases from your breathing zone and general area.
- Keep your head out of fumes.
- Use enough ventilation or exhaust at the arc or both.

ARC RAYS can injure eyes and burn skin.
- Wear correct eye, ear, and body protection.

ENGINE EXHAUST GASES can kill.
- Use in open, well ventilated areas or vent the engine exhaust to the outside.

HIGH VOLTAGE can kill
- Do not touch electrically live parts.
- Stop engine before servicing.

MOVING PARTS can cause serious injury.
- Keep clear of moving parts.
- Do not operate with protective covers, panels, or guards removed.

- Only qualified personnel should install, use, or service this equipment.
- Read operating manual for details.

READ AND UNDERSTAND THE MANUFACTURER'S INSTRUCTIONS AND YOUR EMPLOYER'S SAFETY PRACTICES.

See American National Standard Z49.1, "Safety in Welding and Cutting", published by the American Welding Society, 550 Le Jeune Rd., Miami, Florida 33126; OSHA Safety and Health Standards, 29 CFR 1910, available from U.S. Government Printing Office, Washington, D.C. 20402.

DO NOT REMOVE THIS WARNING E202

Figure 2-13 Note the warning information for electrical shock and high voltage contained on this typical label, which is attached to welding equipment by the manufacturer. (Courtesy of the Lincoln Company)

Eye Protection

You are always at risk of getting particles in your eyes when welding, grinding, sanding, or wire brushing. Adequate eye protection is critical. With any eye injury, the risk of infection is present, so all eye injuries should be considered serious. Protecting your eyes with adequate protective equipment is your first line of defense.

EYE SAFETY ALERT

Infections due to untreated eye injuries are dangerous. Always have eye injuries treated by a physician.

Safety Glasses. Eye safety devices range from basic safety glasses with side shields to protective barriers. Wear safety glasses all the time! Understanding the reasons for this precaution may help you to follow this important practice.

The electric arc, combined with the force of the arc into the weld pool, causes many welding sparks during welding. These sparks are small particles of molten metal, also known as *spatter* when they are large enough to be deposited on the base metal. As sparks fly through the air, some solidify, others remain molten for some time, and others are somewhere in between in a plastic state. Whatever their form, hot particles can cause damage if they hit you or land on your clothing. Protect your eyes from this flying debris with safety glasses and other personal safety equipment.

A deoxidizing agent used to reduce the effects of oxygen in the weld area may produce a glasslike substance that sticks to the surface of the weld bead. This glasslike substance is not harmful to the weld and is easily removed. As the weld bead cools and the metal contracts, the residue does not. This change may cause the residue to pop off. Always wear safety glasses when observing the solidifying weld surface. Safety glasses usually have clear lenses, but some are available in several shades of darkness much like sunglasses. This type of filtered safety eyewear also helps protect the eyes from minor weld flashes (welding light that is reflected or inadvertently seen from a distance). Filtered safety glasses are not to be used for welding without the additional protection of a welder's helmet outfitted with an adequate filter lens. Good safety glasses are equipped with protective side shields and fit snugly around the welder's eye sockets. Filtered safety glasses help if minor weld flashes occur.

CAUTION

Make sure your safety glasses conform to American National Standard Institute (ANSI) code Z87.1 (Table 2-1).

Goggles. In addition to safety glasses, you can also use safety goggles when grinding metals. In this case, wear your safety glasses *under* the goggles for extra protection.

Welding goggles are made of a softer material than safety glasses and form a seal around the safety glasses. This tight fit prevents sparks from bouncing around between the safety glasses and your eye socket.

CAUTION

Use care when removing goggles. The accumulation of dirt and grit buildup on the goggles can fall into your eyes.

Face Shields. Face shields protect the entire face from flying sparks but do not seal off the eye sockets as well as goggles. The advantage of face shields is that the entire face is protected. Face shields are less likely to fog over from body heat and fit more comfortably than goggles. Your best protection from flying sparks is to simultaneously use safety glasses, goggles, and a face shield.

Welding Helmets. The welding helmet is a protective device that keeps the intense waves of light from burning your skin and the hot welding sparks from burning your face and neck. The welding helmet also holds a special glass or plastic lens, called the *filter lens,* which reduces the intensity of the light waves given off by the arc (Figure 2-14).

Filter Lenses. Filter lenses are available in several shades and are numbered for easy identification. The higher the shade number, the darker the filter lens. The higher-numbered lenses allow less light to enter the welding helmet and protect your eyes from the intensity of the arc. The filter lens is usually made of glass. A clear plastic cover protects the filter lens from sparks.

TABLE 2-1

HUNSTMAN SELECTOR CHART. (COURTESY OF KEDMAN CO., HUNSTMAN PRODUCT DIVISION)

Selection Chart for Eye and Face Protectors for Use in Industry and Schools

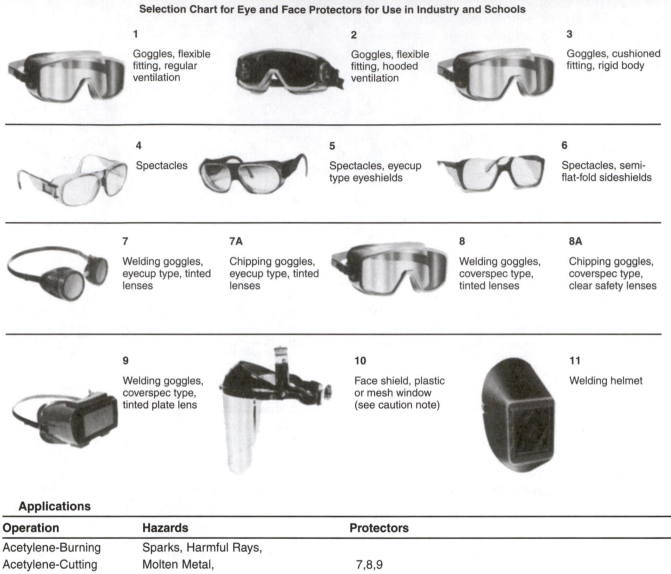

1 Goggles, flexible fitting, regular ventilation

2 Goggles, flexible fitting, hooded ventilation

3 Goggles, cushioned fitting, rigid body

4 Spectacles

5 Spectacles, eyecup type eyeshields

6 Spectacles, semi-flat-fold sideshields

7 Welding goggles, eyecup type, tinted lenses

7A Chipping goggles, eyecup type, tinted lenses

8 Welding goggles, coverspec type, tinted lenses

8A Chipping goggles, coverspec type, clear safety lenses

9 Welding goggles, coverspec type, tinted plate lens

10 Face shield, plastic or mesh window (see caution note)

11 Welding helmet

Applications

Operation	Hazards	Protectors
Acetylene-Burning Acetylene-Cutting Acetylene-Welding	Sparks, Harmful Rays, Molten Metal, Flying Particles	7,8,9
Chemical Handling	Splash, Acid Burns, Fumes	2 (for severe exposure add 10)
Chipping	Flying Particles	1,2,4,5,6,7A,8A
Electric (Arc) Welding	Sparks, Intense Rays, Molten Metal	11 (in combination with 4,5,6 in tinted lenses advisable)
Furnace Operations	Glare, Heat, Molten Metal	7,8,9 (for severe exposure add 10)
Grinding-Light	Flying Particles	1,3,5,6 (for severe exposure add 10)
Grinding-Heavy	Flying Particles	1,3,7A,8A (for severe exposure add 10)
Laboratory	Chemical Splash, Glass Breakage	2 (10 when in combination with 5,6)
Machining	Flying Particles	1,3,5,6 (for severe exposure add 10)
Molten Metals	Heat, Glare, Sparks, Splash	7,8 (10 in combination with 5,6 in tinted lenses)
Spot Welding	Flying Particles, Sparks	1,3,4,5,6 (tinted lenses advisable: for severe exposure add 10)

CAUTION:

Face shields alone do not provide adequate protection. Plastic lenses are advised for protection against molten metal splash.

Contact lenses, of themselves, do not provide eye protection in the industrial sense and shall not be worn in a hazardous environment without appropriate covering safety eyewear.

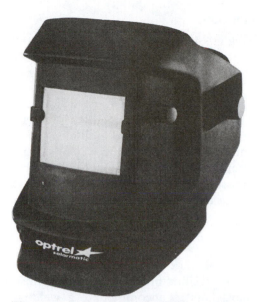

(A)

(B)

Figure 2-14 Typical arc-welding helmets that provide eye and face protection during welding. (Photos courtesy of [top] Thermacote Welco and [bottom] Hornell Speedglas [R. H. Blake])

FILTER LENS SELECTIONS FOR GMAW WELDING

WELDING CURRENT AMPERAGE	FILTER SHADE NO
Under 60	7–10
60–160	10–11
160–250	10–12
250–500	10–14

When selecting a filter lens for an application, first select a higher number (darker) shade. If you find it difficult to see the weld zone, select successively lighter shades until the weld zone is visible. Do not go below the lowest recommended filter lens number.

Selecting the correct filter lens for the task is critical. The American Welding Society has developed a lens selection chart to help you select the proper filter lenses. In addition to the clear and filter lenses, some welders find it useful to use a magnifier lens. This type of lens does not allow you to see any better, but it does make the weld area look larger, giving you the opportunity to control the weld pool better.

An autodarkening welding lens is quite useful. When an arc is struck, some types of glass change automatically to the appropriate darkness at a speed of less than 1/25,000th of a second. A filter in the range from #9 to #13 can be selected with this lens, which is normally set at a shade #5 until the arc is struck. This eliminates guesswork because you see where the weld will start through the shade #5. When the arc is struck, the lens instantaneously darkens to the preselected shade.

CAUTION

Check your welding helmet for holes to prevent your face from being burned during the welding process.

Eyewash Stations. Eyewash stations are areas in many welding shops, factories, and schools that provide immediate, temporary care in the event of a minor eye injury. These can be elaborate systems with piped-in, temperature-controlled water or simply a first-aid station with plastic bottles of sterile eyewash solution. Set up a simple eyewash station in your home hobby shop by keeping bottles of sterile eyewash solution handy. Change these solutions periodically to keep them fresh.

CAUTION

Read directions and know how to use the eyewash station. Post easy-to-read instructions in a visible area near the station.

Flash Protection. Curtains, screens, barriers, or other similar devices can protect workers in the welding area. Flash protection is any device that shields your eyes or exposed skin or protects onlookers from the light produced by a welding arc.

ARC ALERT

See-through screens or curtains are not designed to permit bystanders to view the welding area.

PROTECTIVE CLOTHING

CAUTION

Clothing made from synthetic fibers or clothing that has been soiled by grease or oil should not be worn when welding.

Protective garments are designed for specific tasks where general-use work clothes do not withstand the abusive conditions of welding.

Clothing made from leather or other fire-retardant fabrics, called *welder's leathers,* is designed to prevent the arms, chest, and shoulders from contacting hot metal and exposure to intense welding light.

When ordering welder's leather clothing, ask for a welder's cape, sleeves, vests, bib apron, or welder's coat, all of which are intended for upper-body protection. Lower-body protection is called *welder's leggings.* Welding suppliers can provide you with a catalog of leather clothing of all types.

CAUTION

Keep the top button of your shirt buttoned to prevent weld flashburns to the neck area. Some welders prefer to wear a bandana or neckerchief to prevent such burns.

Welding Gloves

Your hands are subject to damage caused by ultraviolet and infrared light waves, hot metal, and sparks. Most welders prefer soft, all-leather protective gloves with a cloth insulated liner for maximum flexibility and manual dexterity. Uninsulated gloves may be more flexible and comfortable but do not provide as much protection.

SAFETY ALERT

Use metal tongs (pliers) for handling hot metal to prevent burns to the hands when using soft leather gloves.

Welder's Caps

Welder's caps, commonly called *funny fitters' hats,* are a popular and necessary piece of personal protective equipment. Your head is usually under the welder's helmet when welding. The brightly colored caps make you more visible to co-workers.

The welder's cap gives added protection from burns to the top of the head and keeps sparks away from the scalp. The bill of the welder's cap is longer and softer than that of most other types of caps. Your ear is more accessible to welding sparks when you perform tasks requiring that you tilt your head to the side. Turn your cap sideways so its bill covers the ear, and sparks will roll over the bill and away, protecting your ear.

In some trade areas hard hats are required on the job site. Helmet blocks can be easily installed on most hard hats to accommodate a welding helmet (Figure 2-15).

CAUTION

Wear a hard hat when working in areas with overhead hazards.

High-Top Boots

High-top, steel-toe boots with rounded toes and no stitching are recommended when welding and working with metal.

High-top boots fit up over the ankles under the pants' legs, supporting the ankles and preventing sparks from reaching the skin. Sparks are more likely to roll off the surface of rounded-toe boots. Boots without top stitching will last longer.

Figure 2-15 Welding helmet with hard hat. (Courtesy of Kedman Co., Huntsman product division)

NOISE

Noise is unwanted sound. It can be transmitted through gases, liquids, and solids and can physically injure the ears, disrupt relaxation, cause stress, and disrupt sleep patterns. Noise can interrupt concentration and may startle a person to the point of injury by reflex reaction. It can interfere with communication on the job.

When ear damage from noise levels occurs, it may happen so slowly and gradually that you may not even notice it until the damage is done.

The Occupational Safety and Health Administration (OSHA) has published a standard for the minimum level of noise the human ear will withstand without being damaged. Because the measurement of workplace noise is complex, the National Safety Council's "Fundamentals of Industrial Hygiene" and "Noise Control: A Guide for Employees and Employer" are recommended for review to determine noise exposure and remedies for your situation.

Ear Protection

Properly fitted ear protection prevents ear damage from noise caused by welding, grinding, chipping, and cutting metal. In addition, welding sparks can easily enter your unprotected ear canals. Find a type of ear protection that is comfortable, and use it consistently.

HAND TOOLS

Hand tools are used by the auto-body technician to assemble and disassemble parts for repair and welding.

The adjustable wrench is a common tool used by the auto-body technician. When using this wrench, you should adjust it tightly on the nut and push so that most of the force is on the fixed jaw (Figure 2-16). When you are using a wrench on a tight bolt or nut, push the wrench with the palm of your open hand or pull it to prevent injuring the hand. If a nut or bolt is too tight to be loosened with a wrench, obtain a longer wrench. Do not use a "cheater" bar. The fewer points a box end wrench or socket has, the stronger it is and the less likely it is to slip or damage the nut or bolt (Figure 2-17).

Striking a hammer directly against a hard surface such as another hammer face or an anvil may cause chips to fly off and injure someone.

The "mushroomed" heads of chisels, punches, and the faces of hammers should be ground off (Figure 2-18). If you are going to hit a chisel or punch with more than a slight tap, you should hold it in a chisel

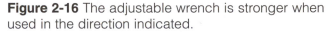
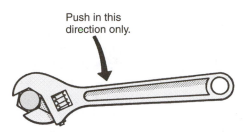

Figure 2-16 The adjustable wrench is stronger when used in the direction indicated.

Figure 2-17 The fewer the points, the less likely the wrench is to slip.

Figure 2-18 Any "mushroomed" heads must be ground off.

holder or with pliers to eliminate the danger of injuring your hand. A handle should be placed on the tang of a file also to avoid injuring your hand (Figure 2-19). A file can be kept free of chips by rubbing a piece of soapstone on it before you use it.

Remember to use the correct tool for the job. Do not try to force a tool to do a job it was not designed to do.

Hand Tool Safety

Several hand tools are used in the auto-body shop. These tools should be treated properly and not abused. Many accidents can be avoided by using the right tool for the job. For instance, a tool that is the correct size for the work should be used instead of one that is too large or too small.

Keep hand tools clean to protect them against corrosion damage. Wipe off any accumulated dirt and grease. Dip the tools occasionally in cleaning fluids or solvents, and wipe them clean. Lubricate adjustable and other moving parts to prevent wear and misalignment.

Make sure that hand-tool cutting edges are sharp. Sharp tools make work easier, improve the accuracy of the work, save time, and are safer than dull tools. When sharpening, redressing, or repairing tools, shape, grind, hone, file, fit, and set them properly using other tools suited to each purpose. For sharpening tools, either an oilstone or a grindstone is preferable. If grinding on an abrasive wheel is required, grind only a small amount at a time with the tool rest set not more than 1/3 inch from the wheel. Hold the tool lightly against the wheel to prevent overheating, and frequently dip the part being ground in water to keep it cool. This will protect the hardness of the metal and help to retain the sharpness of the cutting edge. Tools struck by hammers, such as chisels or punches, should have their heads ground periodically to prevent "mushrooming." Be sure to wear safety goggles when sharpening or redressing tools.

Keep handles secure and safe. Do not rely on friction tape to secure split handles or to prevent handles from splitting. Check wedges and handles frequently. Be sure heads are wedged tightly on handles. Keep handles smooth and free of rough or jagged surfaces. Protect their tips before driving them into tools, or use a proper mallet to avoid splitting or "mushrooming" them. Replace handles that are split, chipped, or unable to be refitted securely.

When swinging any tool, be absolutely certain that no one is within range or can come within range of the swing or be struck by flying material. Always allow plenty of room for arm and body movements and for handling the work. When carrying tools, protect the cutting edges, and carry the tools in such a way that you will not endanger yourself or others. Carry pointed or sharp-edged tools in pouches or holsters. Never create sparks in the presence of flammable materials or explosive vapors.

Hammers and Mallets Safety

The following safety precautions generally apply to all hammers and mallets:

- Check to see that the handle is tight before using hammer or mallet. Never use a hammer or mallet with a loose or damaged handle.

- Always use a hammer or mallet of suitable size and weight for the job.

- Discard or repair any tool if the face shows excessive wear, dents, chips, mushrooming, or improper redressing.

- Rest the face of the hammer on the work before striking to "get the feel" or aim; then grasp the handle firmly near the extreme end of the handle. Get the fingers out of the way before striking with force.

- A hammer blow should always be struck squarely, with the hammer face parallel to the surface being struck. Always avoid glancing blows and over-and-under strikes.

- For striking another tool (cold chisel, punch, wedge, and so on), the face of the hammer should be proportionately larger than the head of the tool. For example, a 1/2-inch cold chisel requires at least a 1-inch hammer face.

- Never use one hammer to strike another hammer.

- Do not use the end of the handle of any tool for tamping or prying; it might split.

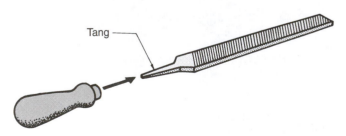

Figure 2-19 To protect yourself from the sharp tang of a file, always use a handle with a file.

POWER TOOLS

All power tools must be properly grounded to prevent accidental electrical shock. If you feel even a slight tingle while using a power tool, stop and have the tool checked by an electrician. Power tools should never be used with force or allowed to overheat from excessive or incorrect use. If an extension cord is used, it should have a large enough current rating to carry the load. An extension cord that is too small will cause the tool to overheat.

Safety glasses must be worn at all times when you use power tools.

Drills

Secure the workpiece as necessary, and fasten it in a vise or clamp. Holding a small item in your hand can cause injury if it is suddenly seized by the bit and whirled from your grip. This is most likely to happen just as the bit breaks through the hole at the backside of the work. All sheet metal tends to cause the bit to grab as it goes through. This can be controlled by reducing the pressure on the drill just as the bit starts to go through the workpiece.

Carefully center the drill bit in the jaws of the chuck and securely tighten. Avoid inserting the bit off-center because it will wobble and probably break when it is used. Drill bits that are 1/4 inch (6 mm) may be hand-tightened in the drill chuck to prevent them from snapping if they accidentally "grab." Hand-tightening the small bits allows them to spin in the chuck if necessary, thus reducing bit breakage. This technique does not always work because some chucks cannot hold the bit securely enough to prevent it from spinning during normal use. In such situations the chuck must be tightened securely with a chuck key.

When possible, center punch the workpiece before drilling to prevent the drill bit from walking across the surface as the drilling begins (Figure 2-20).

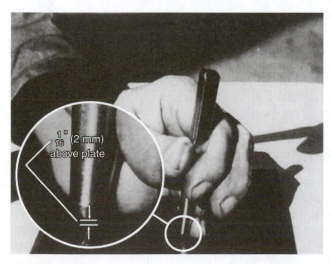

Figure 2-20

After centering the drill bit tip on the exact point at which the hole is to be drilled, start the motor by pulling the trigger switch. Never apply a spinning drill bit to the work. Run a variable-speed drill at a very low speed until the cut has begun. Then, gradually increase to the optimum drill speed.

Except when it is desirable to drill a hole at an angle, hold the drill perpendicular to the face of the work. Align the drill bit and the axis of the drill in the direction the hole is to go, and apply pressure only along this line, with no sideways or bending pressure. Changing the direction of this pressure will distort the dimensions of the hole, which could snap a small drill.

Use just enough steady and even pressure to keep the drill cutting. Guide the drill by leading it slightly, if needed, but do not force it. Too much pressure can cause the bit to break or overheat. Too little pressure will keep the bit from cutting and dull its edges due to the friction created by sliding over the surface.

If the drill becomes jammed in the hole, release the trigger immediately, remove the drill bit from the work, and determine the cause of the stalling or jamming. Do not squeeze the trigger on or off in an attempt to free a stalled or jammed drill. When using a reversing-type model, you may reverse the direction of the rotation to help free a jammed bit. Be sure the direction of the rotation is reset before attempting to continue the drilling.

Reduce the pressure on the drill just before the bit cuts through the work to avoid stalling in metal. When the bit has completely penetrated the work and is spinning freely, withdraw it from the work while the motor is still running, and then turn off the drill.

Grinders

Grinding using a portable or a pedestal grinder is required on most auto-body jobs. Often it is necessary to grind weld, remove rust, or smooth a surface. Grinding stones can make the maximum number of revolutions per minute (r/min) listed on the paper blotter (Figure 2-21). They must never be used on a machine with a higher "r/min" rating. If grinding stones are turned too fast, they can explode.

Grinding Stones

Before a grinding stone is put on the machine, it should be tested for cracks. This is done by tapping the stone in four places and listening for a sharp ring, which indicates it is good (Figure 2-22). A dull sound indicates that the grinding stone is cracked and should not be used. Once a stone has been installed and used,

Figure 2-21 Always check to be sure that the grinding stone and the grinder are compatible before installing a stone.

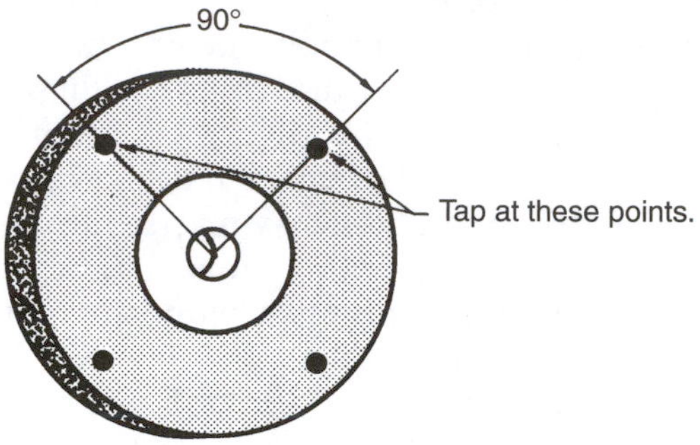

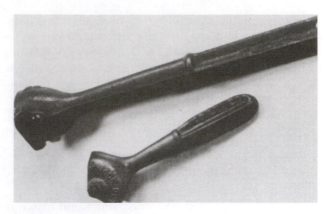

Figure 2-23 Use a grinding stone redressing tool as needed to keep the stone in balance.

Figure 2-22 Tap the stone to check for cracks.

Figure 2-24 Keep the tool rest adjusted.

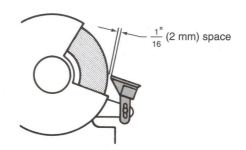

it may need to be trued and balanced by using a special tool designed for that purpose (Figure 2-23). Truing keeps the stone face flat and sharp for better results.

Types of Grinding Stones. Each grinding stone is made for grinding specific types of metal. Most stones are for ferrous metals, for example, iron, cast iron, steel, and stainless steel. Some stones are made for nonferrous metals such as aluminum, copper, and brass. If a ferrous stone is used to grind nonferrous metal, the stone will become glazed (the surface clogs with metal) and may explode due to frictional heat building up on the surface. If a nonferrous stone is used to grind ferrous metal, the stone will be quickly worn away.

When the stone wears down, keep the tool rest adjusted to within 1/16 in. (2 mm) (Figure 2-24), so

that the metal being ground cannot be pulled between the tool rest and the stone surface. Stones should not be used when they are worn down to the size of the paper blotter. If small parts become hot from grinding, pliers can be used to hold them. Gloves should never be worn when grinding. If a glove gets caught in a stone, the whole hand may be drawn in.

The sparks from grinding should be directed down and away from other workers and vehicles and especially from glass. If it is not possible to direct sparks to where they cannot cause damage, the area the sparks are hitting must be covered with a fire-resistant blanket.

EQUIPMENT MAINTENANCE

A routine schedule of equipment maintenance will aid in detecting potential problems such as leaking shielding gas, loose wires, poor grounds, frayed insulation, or split hoses. Small problems, if fixed in time, can prevent the loss of valuable time due to equipment breakdown or injury.

Any maintenance beyond routine external maintenance should be referred to a trained service technician. In most areas, it is against the law for anyone but a licensed electrician to work on equipment such as arc welders and anyone but a factory-trained repair technician to work on regulators. Electrical shock and exploding regulators can cause serious injury or even death.

Hoses

Hoses must be used only for the gas or liquid for which they were designed. Green hoses are to be used only for oxygen, and red hoses are to be used only for acetylene or other fuel gases. Orange or black hoses are used for compressed air. Avoid using unnecessarily long lengths of hoses. Never use oil, grease, lead, or other pipe-fitting compounds for any joints. Hoses should also be kept out of the direct line of sparks. Any leaking or bad joints in gas hoses must be repaired.

WORK AREA

The work area should be kept picked up and swept clean. Collections of scrap, new stock, wire, hoses, and power cables are difficult to work around and easy to trip over. Hooks can be made to hold hoses and cables, and scrap should be thrown into scrap bins.

Portable screens should be used whenever arc welding is to be done in an area where others are working (Figure 2-25).

If you are going to leave a piece of hot metal unattended, write the word *hot* on it before leaving. This procedure can also be used to warn people of hot tables, vises, and tools.

MATERIAL HANDLING

Proper lifting, moving, and handling of large or heavy parts is important to the safety of auto-body technicians. Improper work habits can cause serious personal injury.

Lifting

When you lift a heavy object, the weight of the object should be distributed evenly between both hands, and your legs should do the lifting, not your back (Figure 2-26). Do not try to lift a large or bulky

Figure 2-25 Portable safety screen.

Figure 2-26 Lift with your legs, not your back.

object without help if the object is heavier than you can lift with one hand.

EXTENSION CORDS

If there is some distance from the power source to the work area or if a portable tool is equipped with a stub power cord, you must use an extension cord. When using extension cords on portable power tools, the size of the conductors must be large enough to prevent an excessive drop in voltage. A voltage drop is the lowering of the voltage at the power tool from the voltage at the supply. The reduction of voltage is the result of resistance to electrical flow in the wire. A voltage drop would cause loss of power, overheating,

TABLE 2-2

RECOMMENDED EXTNSION CORD SIZES FOR USE WITH PORTABLE ELECTRIC TOOLS															

NAMEPLATE AMPERE RATING

CORD LENGTH	0 TO 5	6	8	8	9	10	11	12	13	14	15	17	17	18	19	20
25'	18	18	18	18	18	18	16	16	16	14	14	14	14	14	12	12
50'	18	18	18	18	18	18	16	16	16	14	14	14	14	14	12	12
75'	18	18	18	18	18	18	16	16	16	14	14	14	14	14	12	12
100'	18	18	18	16	16	16	16	16	14	14	14	14	14	14	12	12
125'	18	18	16	16	16	14	14	14	14	14	14	14	12	12	12	12
150"	18	16	16	16	14	14	14	14	14	12	12	12	12	12	12	12

Note: Wire sizes shown are AWG (American Wire Gauge) based on a line voltage of 120.

and possible motor damage. Table 2-2 shows the correct size to use based on cord length and nameplate amperage rating. If in doubt, use the next larger size. The smaller the gauge number of an extension cord, the larger the cord.

Two-wire extension cords with two-prong plugs are not acceptable in most auto-body shops. Only three-wire, grounded extension cords connected to properly grounded three-wire receptacles may be used in auto-body shops (Figure 2-27). Current specifications require outdoor receptacles to be protected with ground-fault interpreter (GFI) devices.

When using extension cords, keep in mind the following safety tips:

- Always connect the cord of a portable electric power tool to the extension cord before you connect the extension cord to the outlet. Always unplug the extension cord from the receptacle before you unplug the cord of the portable power tool from the extension cord.

- Extension cords should be long enough to make connections without being pulled taut, creating unnecessary strain and wear.

- Be sure that the extension cord does not come in contact with sharp objects or hot surfaces. Do not allow cords to kink or be dipped in or splattered with oil, grease, or chemicals.

- Before using a cord, inspect it for loose or exposed wires and damaged insulation. If a cord is damaged, it must be replaced. This also applies to the tool's power cord.

- Extension cords should be checked frequently while in use to detect unusual heating. Any

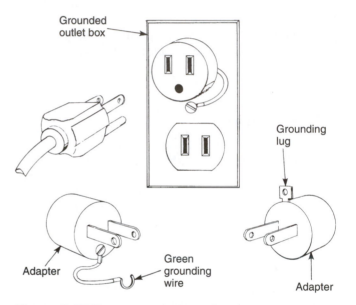

Figure 2-27 Three-prong plug adapters.

cable that feels more than slightly warm to the bare hand placed outside the insulation should be checked immediately for overloading.

- See that the extension cord is placed to prevent tripping or stumbling.

- To prevent the accidental separation of a tool cord from an extension cord during operation, make a knot as shown in Figure 2-28A, or use a cord connector as shown in Figure 2-28B.

- Use an extension cord that is long enough for the job but not excessively long.

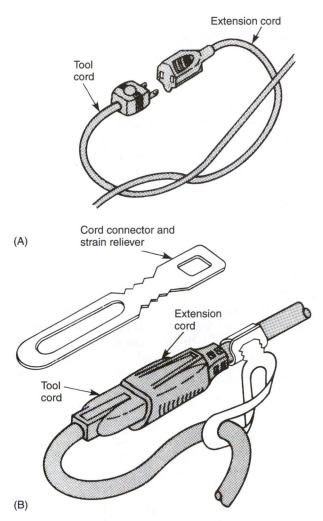

Extension cord

Tool cord

(A)

Cord connector and strain reliever

Extension cord

Tool cord

(B)

Figure 2-28 (A) A knot will prevent the extension cord from accidentally pulling apart from the tool cord during operation. (B) A cord connector will serve the same purpose.

SAFETY RULES FOR PORTABLE ELECTRIC TOOLS

In all tool operation, safety is simply the removal of any element of chance. A few safety precautions that should be observed are given in the following list. These are general rules that apply to all power tools. They should be strictly obeyed to avoid injury to the operator and damage to the power tool.

- Know the tool. Learn its applications and limitations as well as its specific potential hazards by reading the manufacturer's literature.
- *Ground the portable power tool unless it is double-insulated.* If the tool is equipped with a three-prong plug, it must be plugged into a three-hole electrical receptacle. If an adapter is used to accommodate a two-pronged receptacle, the adapter wire must be attached to a known ground. *Never remove the third prong.*
- Do not expose the power tool to rain. Do not use a power tool in wet locations.
- Keep the work area well lighted. Avoid chemical or corrosive environments.
- Because electric tools create sparks, portable electric tools should never be started or run in the presence of propane, natural gas, gasoline, paint thinner, acetylene, or other flammable vapors. To do so would be to risk causing a fire or an explosion.
- *Do not force a tool.* It will do the job better and more safely if operated at the rate for which it was designed.
- Use the right tool for the job. Never use a tool for any purpose except that for which it was designed.
- *Wear eye protectors.* Safety glasses or goggles will protect the eyes while you operate power tools.
- Wear a face or dust mask if the operation creates dust.
- Take care of the power cord. Never carry a tool by its cord or yank it to disconnect it from a receptacle.
- Secure your work. Use clamps to hold the work because this is safer than using your hands, and it frees both hands to operate the tool.
- Do not overreach when operating a power tool. Keep proper footing and balance at all times.
- Maintain power tools. Follow the manufacturer's instructions for lubricating and changing accessories. Replace all worn, broken, or lost parts immediately.
- *Disconnect tools from the power source when not in use.*
- *Remove adjusting keys and wrenches before operation.* Form the habit of checking to see that any keys or wrenches are removed from a tool before turning it on.
- *Avoid accidental starting.* Do not carry a plugged-in tool with your finger on the switch. Be sure the switch is off when you plug in the tool.

- *Be sure accessories and cutting bits are attached securely to the tool.*
- *Do not use tools with cracked or damaged housing.*
- When operating a portable power tool, *give it your full and undivided attention.*

HANDLING AND STORING CYLINDERS

Oxygen and fuel gas cylinders or other flammable materials must be stored separately at a distance of at least 20 feet (6.1 m) or behind a wall 5 feet high (1.5 m) with at least a 1/2-hour burn rating (Figure 2-29). The purpose of the distance or wall is to keep the heat from a small fire from causing the oxygen cylinder safety valve to release. If the safety valve releases the oxygen, a small fire would become a raging inferno.

Inert gas cylinders may be stored separately or with either fuel cylinders or oxygen cylinders.

Empty cylinders must be stored separately from full cylinders, although they may be stored in the same room or area. All cylinders must be stored vertically and have the protective caps screwed on firmly.

Securing Gas Cylinders

Cylinders must be secured with a chain or other device so that they cannot be knocked over accidentally. Even though more stable, cylinders attached to a manifold or stored in a special room used only for cylinder storage should be chained.

Storage Areas

Cylinder storage areas must be located away from halls, stairwells, and exits so that, in an emergency, they will not block an escape route. Storage areas should also be located away from heat, radiators, furnaces, and welding sparks. The location of storage areas should prevent unauthorized people from tampering with the cylinders. A warning sign that reads "Danger—No Smoking, Matches, or

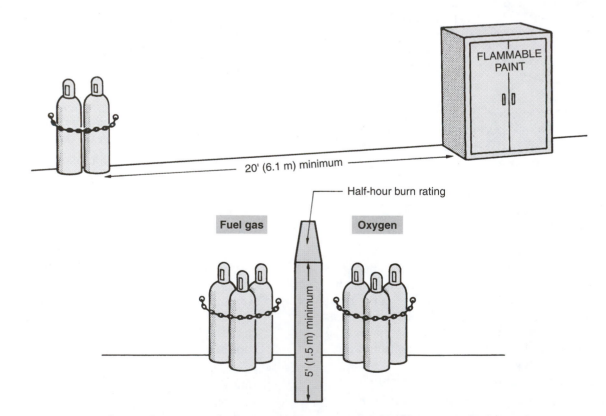

Figure 2-29 Stored fuel gas cylinders must be separated from flammable materials by at least 20 feet (6.1 m) or a wall 5 feet (1.5 mm) high.

Open Lights" should be posted in the storage area (Figure 2-30).

Cylinders with Valve Protection Caps

A cylinder equipped with a valve protection cap must have the cap in place unless the cylinder is in use. The protection cap prevents the valve from being broken off if the cylinder is knocked over. If the valve of a full, high-pressure cylinder of argon, oxygen, carbon dioxide (CO_2), and mixed gases is broken off, the cylinder valve will be propelled around the shop like a missile if it has not been secured properly. Never lift a cylinder by the safety cap or the valve. The valve can easily break off or be damaged.

When moving cylinders, you should replace the valve protection cap, especially if the cylinders are mounted on a truck or trailer for out-of-shop work. The cylinders must never be dropped or handled roughly.

General Precautions

Use warm water, not boiling, to loosen cylinders that are frozen to the ground. Any cylinder that leaks, has a bad valve, or has gas-damaged threads must be identified and reported to the supplier. Use a piece of soapstone to write the problem on the cylinder. If the leak cannot be stopped by closing the cylinder valve, move the cylinder to a vacant lot or open area. Then slowly release the pressure after posting a warning sign (Figure 2-31).

Acetylene cylinders that have been lying on their sides must stand upright for 15 minutes or more before you use them. The acetylene is absorbed in acetone, and the acetone is absorbed in a filler. The filler does not allow the liquid to settle back away from the valve very quickly (Figure 2-32). If the cylinder has been in a horizontal position, using it too soon after it is placed in a vertical position may draw acetone out of the cylinder. Acetone lowers the flame temperature and can damage regulator or torch valve settings.

Figure 2-31 Move a leaking fuel gas cylinder out of the building or work area. Slowly release the pressure after posting a warning of the danger.

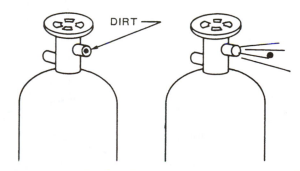

Figure 2-32 During transportation or storage, dirt may collect in the valve. Wiping this dirt may leave oil from a hand or rag and might not remove all the dirt. Cracking the valve is the best way to remove any dirt.

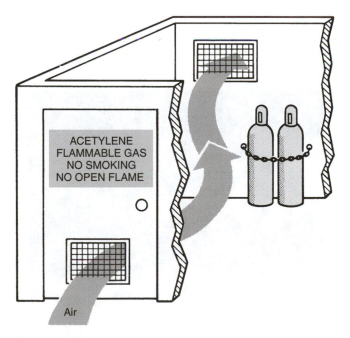

Figure 2-30 Acetylene stored in a separate room must have good ventilation, and the room should have a warning sign on the door.

MATERIALS SPECIFICATIONS DATA SHEETS (MSDSs)

All manufacturers of potentially hazardous materials must provide the users of their products with detailed information regarding possible hazards resulting from the use of their products. These materials specifications data sheets are often called MSDSs. They must be provided to anyone using the product or working in the area where the products are in use. Often companies will post these sheets on a bulletin board or put them in a convenient place near the work area.

VENTILATION

The actual welding area should be well ventilated. Excessive fumes, ozone, or smoke may collect in the welding area. Ventilation should be provided for their removal. Natural ventilation is best, but forced ventilation may be required. Areas that have 10,000 cubic feet (283 cubic meters) or more per welder or that have ceilings 16 feet (4.9 meters) high or higher (Figure 2-33), may not require forced ventilation unless fumes or smoke begin to collect.

Forced Ventilation

Small shops or shops with large numbers of welders require forced ventilation, which can be general or localized using fixed or flexible exhaust pickups (Figure 2-34). General room ventilation must be at a rate of 2,000 cu. ft. (56.4m³) or more per person welding. Localized exhaust pickups must have a suction strong enough to provide 100 linear feet (30.5 m) per minute velocity of welding fumes away from the welder. Local, state, or federal regulations may require that welding fumes be treated to remove hazardous components before they are released into the atmosphere.

Any ventilation system should draw the fumes or smoke away before they rise past the level of the welder's face.

Forced ventilation is always required when welding on metals that contain zinc, lead, beryllium, cadmium, mercury, copper, Austenitic manganese, or other materials that exude dangerous fumes.

HAZARDOUS WASTE

The welding industry uses many types of materials that may be dangerous to the environment and people working in and around welding. These materials are called *hazardous*. Some hazardous materials are fluxes from discarded welding rod subs, paints and thinners, cardboard boxes, styrofoam packaging "peanuts" and packing paper, grinding grit, metal chips of various types, wiping rags, expired spray paint and dye penetrant cans, petroleum oils, grease, and solvents.

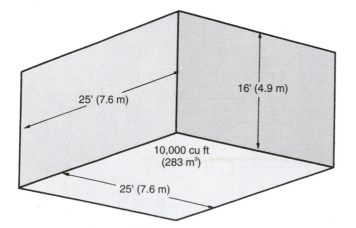

Figure 2-33 A room with a ceiling 16 feet (4.9 m) high may not require forced ventilation for one welder.

Figure 2-34 A flexible exhaust pickup.

These hazardous materials must be disposed of safely while complying with all local, state, and federal regulations. Consult local authorities and MSDs to determine the proper way to handle the waste generated by welding activities.

HORSEPLAY

Horseplay, running, and otherwise "fooling around" should not be tolerated in any shop. Horseplay is also unprofessional and wastes time and can result in your being fired.

SUMMARY

The safety of the welder working in the shop is of utmost importance to the industry. A sizable amount of money is spent on the protection of workers. Some large shops may have an employee (usually a foreperson specifically in charge of shop safety. This person's, job is to ensure that all welders comply with safety rules during production. The proper clothing, shoes, and eye protection to be worn are emphasized in these shops. Any worker who does not follow established safety rules is subject to dismissal.

If an accident does occur, appropriate and immediate first-aid steps must be taken. All welding shops should have established plans for accidents. You should take time to learn the proper procedure for accident response and reporting *before* you need to respond in an emergency. After the situation has been properly taken care of, you should fill out an accident report.

Equipment is periodically checked to be sure that it is safe and in proper working condition. Maintenance workers are sometimes employed to ensure that the equipment is in proper working condition at all times.

Further safety information is available in *Safety for Welders* by Larry F. Jeffus (published by Delmar Publishers); from the American Welding Society; and from the U.S. Department of Labor (OSHA) Regulations.

REVIEW QUESTIONS

1. What are the three types of light?
2. Which type of light is the most dangerous?
3. What are the four types of fire extinguishers?
4. What are good safety glasses equipped with?
5. What is the advantage of face shields?
6. What should be done if a slight tingle is felt while using a power tool?
7. Before a grinding stone is put on a machine, it must first be tested for what?
8. Should gloves be worn during grinding?
9. What are green hoses used for?
10. When should portable screens be used?
11. What is a voltage drop?
12. What type of extension cord must be used in an auto-body shop?
13. Acetylene cylinders that have been lying on their sides must stand upright for how long before they are used?
14. What are MSDSs?
15. When is forced ventilation required?

CHAPTER 3

POWER SUPPLIES AND CURRENTS

EQUIPMENT

The basic GMA welding equipment consists of the gun, electrode (wire) feed unit, electrode (wire) supply, power source, shielding gas supply with flowmeter/regulator, control circuit, and related hoses, liners, and cables (Figures 3-1 and 3-2). The system should be portable, and in some cases, it can be used for more than one process. These power sources can be switched over for other uses.

Power Source

The power source has a transformer and a rectifier that produce a DC welding current of 40–600 amperes with 10–40 volts, depending upon the machine. In the past, some GMA welding processes used AC welding current, but (DCRP) is used exclusively for all GMA work.

Because of the long periods of continuous use, GMA welding machines have a 100% duty cycle, which allows the machine to be run continuously without damage.

WELDING MACHINES

To better understand the terms that describe the different welding power supplies, you need to know the following electrical terms:

- *Voltage* or *volts* (V) is a measurement of electrical pressure, in the same way that pounds per square inch is a measurement of water pressure.

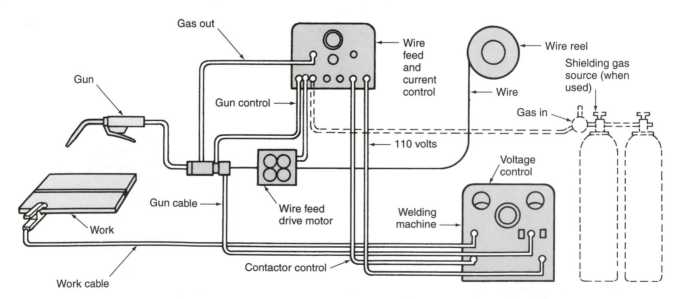

Figure 3-1 Schematic of equipment setup for GMA welding. (Courtesy of Hobart Brothers Company, Troy, Ohio)

- *Electrical potential* means the same thing as voltage and is usually expressed by the term *potential* (P). The terms *voltage, volts,* and *potential* can all be interchanged when referring to electrical pressure.

- *Amperage* or *amps* (A) is the measurement of the total number of electrons flowing, in the same way that gallons is a measurement of an amount of water flowing.

- *Electrical current* means the same thing as amperage and is usually expressed by the term

current (C). The terms *amperage, amps,* and *current* can all be interchanged when referring to electrical flow.

GMA welding power supplies are constant-voltage, constant-potential- (CV, CP) type machines, unlike stick arc-welding power supplies, which are the constant-current (CC) type. It is impossible to make acceptable welds using the wrong type of power supply. Constant-voltage power supplies are available as transformer-rectifiers or as motor-generators (Figure 3-3). Some newer machines use

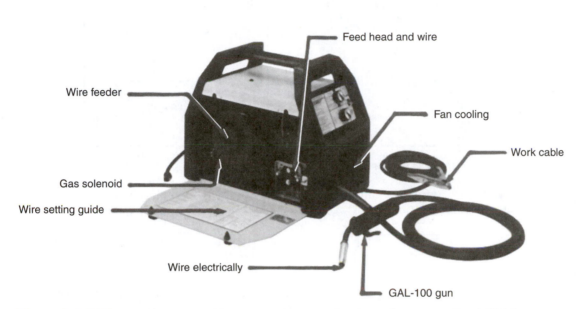

Figure 3-2 GMA welding setup. (Courtesy of Hobart Brothers Company, Troy, Ohio)

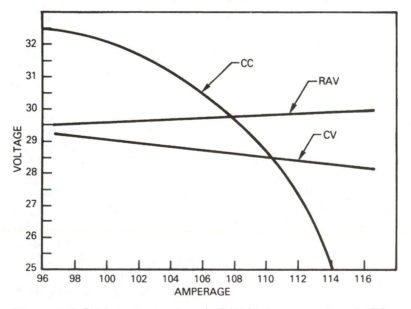

Figure 3-3 CV (constant voltage), RAV (rising arc voltage), CC (constant current).

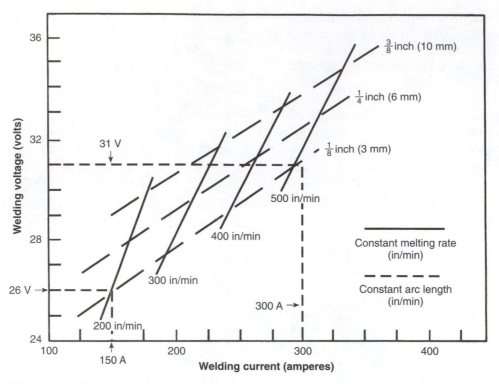

Figure 3-4 The arc length and arc voltage are affected by the welding current and wire-feed speed. (0.045-inch [1.43 mm] wire; one-inch (25 mm) electrode extension)

electronics to enable them to supply both types of welding power by simply flipping a switch.

The relationships between current and voltage with different combinations of arc length or wire feed speeds are called volt–ampere characteristics. The volt–ampere characteristics of arcs in argon with constant arc-lengths or constant wire feed speeds are shown in Figure 3-4. To maintain a constant arc-length while increasing current, it is necessary to increase voltage. For example, with a 1/8-inch arc-length, increasing current from 150 to 300 amperes requires a voltage increase from about 26 to 31 volts. The current increase illustrated here results from increasing the wire feed speed from 200 to 500 inches per minute.

By using Table 3-1, you can determine the amperage by knowing the wire feed speed. Conversely, the wire feed speed can be determined by knowing the voltage. This relationship of amperage to wire feed speed results from the need for a higher amperage to melt the larger quantity of filler metal being introduced to the arc.

Power Supplies for Short-Circuiting Transfer

Although the GMA welding power source is said to have a constant potential (CP), it is not perfectly constant. The graph in Figure 3-5 shows that there is a slight decrease in voltage as the amperage increases within the working range. The rate of decrease is known as *slope* and is expressed as the voltage decrease per 100-ampere increase (for example, 10 V/100 A). For short-circuiting welding, some welding machines are equipped to allow changes in the slope by steps or continuous adjustment.

The slope, which is called the *volt–ampere curve,* is often drawn as a straight line because it is fairly straight within the working range of the machine. Regardless of whether it is drawn as a curve or a straight line, the slope can be found by finding two points. The first point is the set voltage as read from the voltmeter when the gun switch is activated but no welding is being done. This is referred to as the *open-circuit voltage.* The second point is the voltage and amperage as read during a weld. The voltage control is not adjusted during the test, but the amper-

TABLE 3-1

TYPICAL AMPERAGES FOR CARBON STEEL				
WIRE FEED SPEED* IN/MIN (M/MIN)	WIRE DIAMETER			
	.030 IN (0.8 MM)	.035 IN (0.9 MM)	.045 IN (1.2 MM)	.062 IN (1.6 MM)
100 (2.5)	40	65	120	190
200 (5.0)	80	120	200	330
300 (7.6)	130	170	260	425
400 (10.2)	160	210	320	490
500 (12.7)	180	245	365	—
600 (15.2)	200	265	400	—
700 (17.8)	215	280	430	—

*To check feed speed, run out wire for 1 minute and then measure its length.

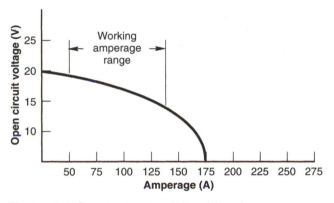

Figure 3-5 Constant-potential welder slope.

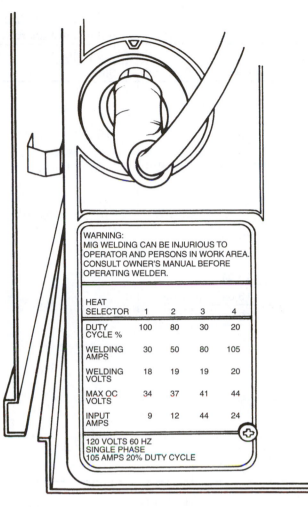

Figure 3-6 Duty-cycle chart.

age can be changed. The slope is the voltage difference between the first and second readings and can be found by subtracting the second voltage from the first voltage. Therefore, for settings over 100 amperes, it is easier to calculate the slope by adjusting the wire feed so that you are welding with 100 amperes, 200 amperes, 300 amperes, and so on. In other words, the voltage difference can be simply divided by 1 for 100 amperes, 2 for 200 amperes, and so forth.

The duty cycle of a particular machine indicates how many minutes out of 10 a welding machine can run without overheating. Figure 3-6 shows the duty cycle chart for a welder. When the heat setting is 30 amps, this machine has a 100% duty cycle, so it can weld 10 minutes out of 10 minutes. The same

machine has only a 20% duty cycle when the heat setting is 105 amps, so it is safe to weld only 2 minutes out of 10.

All machines should have a duty cycle displayed either in the owner's manual or on the machine. The higher the duty cycle, the more expensive the machine. Very little auto-body welding requires long periods of continuous welding; therefore, a typical auto-body welder does not need to have a 100% duty cycle rating.

Some machines also display a welding guide (Figure 3-7). A typical chart identifies the type of base metal, thickness, and the recommended machine settings for solid wires and small-diameter, flux-cored wires. This type of chart is a great convenience when setting the machine.

WELDING GUIDE
Settings are approximate. Adjust as required.

Material	Thickness	Process	Wire		Gas		Polarity	Stickout	Welding Voltage	Wire Speed Control
			Class	Size	Type	Flow				
Carbon Steel	24 ga	GMAW	ER-70S-6 (HB 28)	0.024	CO_2 or C_{25}	20 CFH	DCEP	1/4	1	5.5–6
	18 ga							1/4–5/16	1	6–7
	16 ga							5/16–1/2	2	6.5–7
	10 ga							5/16–1/2	3 (4)	7 (8)
	3/16"							1/2	4	8
	18 ga			0.030				5/16	2	5.5–6.5
	16 ga							5/16–1/2	2	6–6.5
	10 ga							1/2	3	6.5
	3/16"							1/2	4	7–7.75
	18 ga			0.035				5/16	2	5–5.5
	16 ga							5/16	2	5.5
	10 ga							1/2	3	6
	3/16"							1/2	4	6.5
Stainless Steel	10 ga		ER-308L 308L Stainless	0.030	C_{25}			1/2	4	7.5
Carbon Steel	18 ga	FCAW	E-717-11 Fabshield 21B	0.045	None		DCEN	1/2–3/4	2	5
	16 ga							1/2–3/4	2	5.5
	10 ga							3/4	3	6
	3/16"							3/4	4	6

Figure 3-7 Typical welding guide.

Wire-Feed Unit

The purpose of the wire feeder is to provide a steady and reliable supply of wire to the weld. Slight changes in the rate at which the wire is fed have distinct effects on the weld.

The motor of a feed unit can be continuously adjusted over the desired range. There are three major types of wire feed systems: push-type, pull-type, and spool gun-type.

Push-Type Feed System

The wire rollers are clamped securely against the wire to provide the necessary friction to push the wire through the conduit to the gun. The pressure applied on the wire can be adjusted. A groove is provided in the roller to aid in alignment and to lessen the chance of slippage. Most manufacturers provide rollers with smooth or knurled U-shaped or V-shaped grooves (Figure 3-8). Knurling (a series of ridges cut into the groove) helps grip larger-diameter wires so that they can be pushed along more easily. Soft wires, such as aluminum, are easy to damage if knurled rollers are used. Aluminum wires work best with U-grooved rollers. Even V-grooved rollers can distort the surface of soft wire, causing problems. For hard wires, such as mild steel and stainless steel, V-grooved rollers are best. It is also important to use the correct size grooves in the rollers.

In the push-type system, the electrode must have enough strength to be pushed through the conduit without kinking. Mild steel and stainless steel can be readily pushed 15–20 feet (4–6 m), but aluminum is much harder to push over 10 feet (3 m).

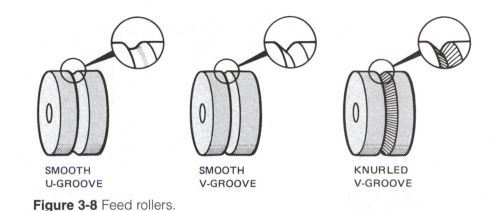

SMOOTH
U-GROOVE

SMOOTH
V-GROOVE

KNURLED
V-GROOVE

Figure 3-8 Feed rollers.

Pull-Type Feed System

In pull-type systems, a smaller but higher-speed motor is located in the gun to pull the wire through the conduit. Using this system, it is possible to move even soft wire over great distances. The disadvantages are that the gun is heavier and more difficult to use, rethreading the wire takes more time, and the operating life of the motor is shorter.

Spool Gun-Type

A spool gun is a compact, self-contained system consisting of a small drive system and a wire supply (Figure 3-9). This system allows the welder to move freely around a job with only a power lead and shielding gas hose to manage. The major control system is usually mounted on the welder. The feed rollers and motor are found in the gun just behind the nozzle and contact tube. Because of the short distance the wire must be moved, very soft wires (aluminum) can be used. A small spool of welding wire is located just behind the feed rollers. The small spools of wire required in these guns are often very expensive. Although the guns are small, they feel heavy when being used.

GMAW Wire-Feed Unit

Push-type wire-feed units are the most commonly used. These units are basically all the same. We will study a few main controls that will familiarize you with most wire-feed units.

The wire-feed unit is an important component of GMAW and can be broken down into several parts. You can easily correct malfunctions of the wire feeder once you know the functions of each of its components (see Figures 3-10 and 3-11).

Figure 3-9 Spool gun for GMA welding. (Courtesy of Miller Electric Mfg. Co.)

Components of the Wire Feeder

Drive Motor. The drive motor provides power to the drive roll that propels the electrode wire to the base metal. Your only control of it is turning it on or off and adjusting the motor's speed.

Drive Rolls. The drive rolls are generally a pair of two or a set of four. They are available in configurations such as smooth U-groove rolls, smooth rolls, V-groove rolls, and knurled rolls. Some are designed for use with only one-diameter wire, whereas others are reversible and allow different diameter wires. Other drive rolls can be adjusted for different sizes of wire by inserting a spacer to increase or decrease the

(A)

(B)

Figure 3-10 (A) A 90-ampere power supply with built-in wire feeder for welding sheet steel with carbon dioxide shielding. (B) Modern wire feeder with digital preset and readout of wire-feed speed and closed-loop control. (Courtesy of Miller Electric Mfg. Co.)

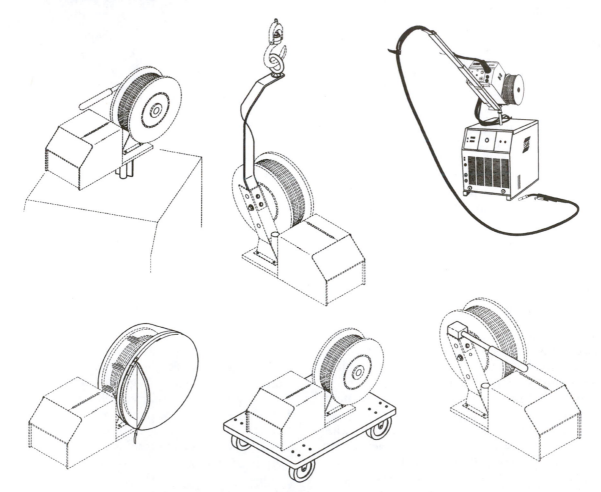

Figure 3-11 A variety of accessories is available for most electrode-feed systems: (A) swivel post, (B) boom hanging bracket, (C) counterbalance miniboom, (D) spool feeder, (E) wire-feeder wheel cart, (F) carrying handle. (Courtesy of ESAB Welding and Cutting)

opening between the rollers. Some rollers are removed and replaced for different wire sizes.

When the rollers are located together, they are called "drive rollers," but when they are separated, they have two different names because they perform different tasks.

- The *drive roll* is connected either directly or indirectly to the drive motor providing the force to push or pull the electrode to the desired location.

- The *tension* or *pressure roller* is not connected to the drive mechanism. It is incorporated into the drive mechanism by an adjustable spring or a combination of a spring and a screw device. The tension rollers (pressure rollers) look just like the drive rollers but are not connected to the wire-feed unit's drive motor, the electric motor that powers the drive roll. The two rollers slightly pinch the electrode, providing tension, and force the wire electrode on its way to the GMAW gun (Figure 3-12).

Tension Adjustment. Tension is maintained on the electrode wire by any of the following:

- A wheel on a screw rotated with the thumb

- A twist screw with a plastic handle operated between the forefinger and thumb

- A spring providing natural, continuous tension with no manual adjustment.

Wire Guides. The steel, copper, brass, or plastic wire guides are such a simple component of the feeder system that it is easy to forget about them. They allow the wire to smoothly enter and exit the drive rolls.

If the wire guides become blocked, the wire is restricted, or the welding variables are incorrect, the wire speed is affected, allowing the electrode to burn back to the contact tube. Wire guides wear because the steel wire continuously running through the softer guides causes grooves, or slotting, in them.

Electrode Packaging. The packaging device that holds the wire electrode is called the *coil, reel, spool,* or *drum*. It is mounted on a spindle, or axle, allowing the wire to easily unroll from its holder (Figure 3-13).

GMAW Guns

There are many brands of welding guns. Figure 3-7 shows an exploded view of a GMAW gun so you can identify the various components of the guns.

Hold the main part of the gun (called the *gun body*) in your hand. The gun trigger is attached to the gun body on most newer guns (Figure 3-14). The

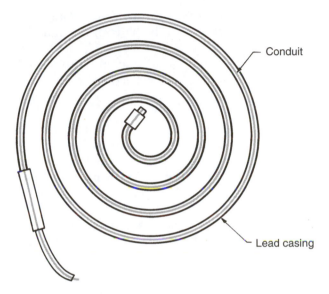

Figure 3-13 Tightly coiled lead casing will force the liner out of the gun.

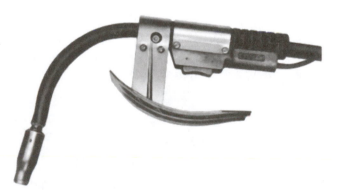

Figure 3-14 A typical GMA welding gun used for most welding processes with a heat shield attached to protect the welder's glove hand from intense heat generated when welding with high amperages. (Courtesy of the Lincoln Electric Company)

Figure 3-12 Adjust the wire-feed tensioner.

trigger closes a switch when depressed, making electrical contact. This activates the drive roll and begins the flow of wire to the contact tube.

The GMAW gun trigger is a contact switch that provides power to the gun. Depress the trigger to make contact, and release it to break contact. Using the trigger allows you to set the electrode wire where you want it without arcing. Then you can position your welding helmet. As long as the trigger is not depressed, the arc will not be struck.

Some older GMAW guns have hot wires that must be "scratch-started" to begin the flow of electrode wire. They do not operate with a switch, so as soon as the electrode wire makes contact with the base metal, the arc is visible, and the wire begins to flow. The welder must have the welding helmet in place before scratch-starting the electrode.

The insulated conductor tube, sometimes called the *gun neck,* is attached to the body of the gun. The conduit liner is a flexible hollow tube that is positioned through the body of the gun and into the gun conductor tube and allows the electrode to snake its way from the feeder unit to the weld pool area (called the *weld zone*).

Replacement conduit liners can be ordered to fit individual guns, or they may be cut to length to fit some gun leads from the feed unit to the GMAW gun. When the conduit liner is in place through the lead, the gun body, and the insulated gun conductor tube, it contacts the gun contact tube. This contact tube allows the electrode to slide through it, making contact. The tube will be electrically hot when the trigger is depressed. The contact tube is screwed into a gas diffuser that is also electrically hot when the trigger is depressed. The gas diffuser allows the shielding gas to be dispersed or diffused around the contact tube, enveloping the entire weld area for greater protection. To keep the electrically hot component from accidentally arcing, an insulating device called the *gun nozzle insulator* is used. This screws onto the conductor tube (neck) and keeps the nozzle away from the contact tube. The nozzle directs the shielding gas around the molten weld pool.

ARC–VOLTAGE AND AMPERAGE CHARACTERISTICS

The arc–voltage and amperage characteristics of GMA welding are different from most other welding processes. The voltage is set on the welder, and the amperage is set by changing the wire-feed speed. At any one voltage setting, the amperage required to melt the wire as it is fed into the weld must change. The faster the wire is fed, the more amperage is required to melt the wire. The slower the wire is fed, the less amperage is required to melt the wire.

Because changes in the wire-feed speed directly change the amperage, it is possible to set the amperage by using a chart and measuring the length of wire fed per minute (Table 3-1). The voltage and amperage required for a specific metal transfer method will be different for various wire sizes, shielding gases, and metals.

The voltage and amperage setting will be specified for all welding that is done according to a welding procedure specification (WPS) or other codes and standards. However, most welding such as that done in small production shops, maintenance welding, repair work, farm shops, and so on is not done to a specific code or standard, and therefore no specific settings exist. For that reason it is important that you learn to make the necessary adjustments to allow you to produce quality welds.

EXPERIMENT 3-1   **SETTING THE CURRENT**

Using a properly assembled GMA welding machine, proper safety protection, and one piece of mild steel plate approximately 12 in. (305 mm) long × 1/4 in. (6 mm) thick, you will change the current settings and observe the effect on GMAW.

On a scale of 0 to 10, set the wire-feed speed control dial at 5, or halfway between the low and high settings of the unit. Also set the voltage at a point halfway between the low and high settings. The shielding gas can be carbon dioxide (CO_2), argon, or a mixture of the two. The gas flow should be adjusted to a rate of 35 cf/h (16 L/min).

Hold the welding gun at a comfortable angle, lower your welding hood, and pull the trigger. As the wire feeds and contacts the plate, the weld will begin. Move the gun slowly along the plate. Note the following welding conditions as the weld progresses: voltage, amperage, weld direction, metal transfer, spatter, molten weld pool size, and penetration. Stop and record your observations in Table 3-2. Evaluate the quality of the weld as "acceptable" or "unacceptable."

Reduce the voltage somewhat and make another weld, keeping all other weld variables (travel speed, stickout, direction, and amperage) the same. Observe the weld, and, upon stopping, record the results. Repeat this procedure until the voltage has been lowered to the minimum value indicated on the

TABLE 3-2

				MOLTEN POOL	
WELD ACCEPTABILITY	**VOLTAGE**	**AMPERAGE**	**SPATTER**	**SIZE**	**PENETRATION**
Good	20	75	Light	Small	Little

SETTING THE CURRENT

Electrode Diameter	.035 in (0.9 mm)
Shielding Gas	CO2
Welding Direction	Backhand

TABLE 3-3

ELECTRODE EXTENSION

WELD ACCEPTABILITY	**VOLTAGE**	**AMPERAGE**	**ELECTRODE EXTENSION**	**CONTACT TUBE-TO-WORK DISTANCE**	**BEAD SHAPE**
Poor	20	100	1 in (25 mm)	1 1/4 in (31 mm)	Narrow, high with little penetration

Electrode Diameter	.035 in (0.9 mm)
Shielding Gas	CO2
Welding Direction	Forehand

machine. Near the lower end the wire may stick, jump, or simply no longer weld.

Return the voltage indicator to the original starting position and make a short test weld. Stop and compare the results to those first observed. Then slightly increase the voltage setting and make another weld. Repeat the procedure of observing and recording the results as the voltage is increased in steps until the maximum machine capability is obtained. Near the maximum setting the spatter may become excessive if CO_2 shielding gas is used. Care must be taken to prevent the wire from fusing to the contact tube.

Return the voltage indicator again to the original starting position, and make a short test weld. Compare the results observed with those previously obtained.

Lower the wire-feed speed setting slightly and use the same procedure as before. First lower and then raise the voltage through a complete range and record your observations. After a complete set of test results is obtained from this amperage setting, again lower the wire-feed speed for a new series of tests. Repeat this procedure until the amperage is at the minimum setting shown on the machine. At low amperages and high-voltage settings, the wire may tend to pop violently as a result of the uncontrolled arc.

Return the wire-feed speed and voltages to the original settings. Make a test weld and compare the results with the original tests. Slightly raise the wire speed, and again run a set of tests as the voltage is changed in small steps. After each series, return the voltage setting to the starting point, and increase the wire-feed speed. Make a new set of tests.

All of the test data can be gathered into an operational graph for the machine, wire type, size, and shielding gas. Use Table 3-3 to plot the graph. The acceptable welds should be marked on the lines that extend from the appropriate voltages and amperages. Upon completion, the graph will give you the optimum settings for the operation of this particular GMAW setup. The optimum settings are found along a line in the center of the acceptable welds.

Experienced welders will follow a much shorter version of this type of procedure any time they start to work on a new machine or test for a new job. This experiment can be repeated using different types of wire, wire sizes, shielding gases, and weld directions. Finally, turn off the welding machine and shielding gas, and clean up your work area.

REVIEW QUESTIONS

1. Electrical potential means the same thing as _____ and is usually expressed by the term *potential* (P).
2. What other two terms can be interchanged with *current* when referring to electrical flow?
3. The relationships between current and voltage with different combinations of arc length or wire-feed speeds are called what?
4. What does the duty cycle of a particular machine indicate?
5. What is the purpose of the wire feeder?
6. What are the three major types of wire-feed system?
7. What wires are best used with U-grooved rollers?
8. What are the disadvantages of the pull-type feed system?
9. What is the drive motor?
10. What is the difference between tension rollers and drive rollers?

CHAPTER 4

SHIELDING GAS AND FLOW METERS

Shielding gases in the gas metal-arc process are used primarily to protect the molten metal from oxidation and contamination. As the weld progresses, a gaseous shield is directed into the weld zone through the gun nozzle to protect the weld pool from atmospheric contamination until the molten weld pool solidifies.

The shielding gas can be provided from a compressed gas cylinder or from a central gas piping system, often referred to as a *manifold system*. Individual cylinders provide the greatest portability, whereas the manifold system offers the greatest potential savings.

The specific shielding gas or gas mixtures used significantly influence the GMA weld produced. The most commonly used gases are carbon dioxide (CO_2), argon (Ar), and helium (H). Sometimes a mixture of these gases is used to obtain the best possible welding performance. And occasionally a trace of oxygen (O) can be added to the mixture for making welds on some ferrous (steel) alloys.

The most commonly used gases for ferrous metals are CO_2, argon with 2-5% oxygen added, and argon with 25% CO_2 added. Nonferrous metals must be welded using inert gases such as argon, helium, or mixtures of argon and helium.

GMAW SHIELDING GASES

Codes and standards often dictate which shielding gas or mixture of gases may be used for specific welds. Even when using such specifications, the welder may still have to choose the specific shielding gas within the ranges allowed by the specification. To select the best shielding gas, you must consider many factors (Table 4-1). The major factors include the following:

- Metal-transfer method
- Weld bead shape; penetration and width of fusion zone
- Welding speed
- Weld discontinuities
- Weld spatter
- Metal transfer efficiency
- Type of base and filler metals
- Welding position
- Cost of the gas
- Total welding cost

Argon

Argon is an inert gas that is a by-product in air separation plants. Air is cooled to temperatures that cause it to liquify; then its constituents are fractionally distilled. The primary products are oxygen and nitrogen. Before these gases were produced on a tonnage scale, argon was a rare gas. Now it is distributed in cylinders as gas or in bulk as liquid forms.

Because argon is denser than air, it effectively shields welds in deep grooves in the flat position.

TABLE 4-1

ADJUSTMENTS IN WELDING VARIABLES AND TECHNIQUES

Welding Variables to Change	Desired Changes							
	Penetration		Deposition Rate		Bead Size		Bead Width	
	Increase	Decrease	Increase	Decrease	Increase	Decrease	Increase	Decrease
Current and Wire Feed Speed	Increase	Decrease	Increase	Decrease	Increase	Decrease	No effect	No effect
Voltage	Little effect	Little effect	No effect	No effect	No effect	No effect	Increase	Decrease
Travel Speed	Little effect	Little effect	No effect	No effect	Decrease	Increase	Increase	Decrease
Stickout	Decrease	Increase	Increase	Decrease	Increase	Decrease	Decrease	Increase
Wire Diameter	Decrease	Increase	Decrease	Increase	No effect	No effect	No effect	No effect
Shield Gas Percent CO2	Increase	Decrease	No effect	No effect	No effect	No effect	Increase	Decrease
Torch Angle	Backhand to 25°	Forehand	No effect	No effect	No effect	No effect	Backhand	Forehand

However, this higher density can be a hindrance when welding overhead because higher flow rates are necessary. Because argon is relatively easy to ionize, this property permits fairly long arcs at lower voltages, making it virtually insensitive to changes in arc length.

Inert gases such as argon provide the necessary shielding because inert gases do not form compounds with any other substance and are insoluble in molten metal. When used as pure gases for welding ferrous metals, argon may produce an erratic arc action and promote undercutting and other flaws. For this reason 100% argon is not normally used for making welds on ferrous metals but is used on nonferrous metals such as aluminum and nickel.

Argon Gas Mixes

Because of argon's problems with ferrous metal welds, it is usually necessary to add controlled quantities of reactive gases to achieve good arc action and metal transfer with these materials. Adding oxygen or carbon dioxide to argon tends to stabilize the arc, promote favorable metal transfer, and minimize spatter. As a result, the penetration pattern is improved, and undercutting is reduced or eliminated.

The amount of the reactive gases, oxygen or carbon dioxide, required to produce the desired effects is quite small. As little as 0.5% of oxygen will produce noticeable change. A more common amount is 1–5% oxygen. Carbon dioxide may be added to argon in the range of 20–30%. Mixtures of argon with less than 10% carbon dioxide may not have enough arc voltage to give the desired results.

Adding oxygen or carbon dioxide to argon causes the shielding gas to oxidize. This in turn may cause porosity in some ferrous metals. In this case, a filler wire containing suitable deoxidizers should be used. The presence of oxygen in the shielding gas can also cause some loss of certain alloying elements, such as chromium, vanadium, aluminum, titanium, manganese, and silicon. Again, the addition of a deoxidizer to the filler wire is necessary.

Helium

Helium is an inert gas that is a by-product of the natural gas industry. It is removed from natural gas as the gas undergoes separation (fractionation) for purification or refinement.

Helium has the disadvantage of being lighter than air; thus its flow rates must be about twice as high as argon's for acceptable stiffness in the gas stream, and proper protection is difficult in drafts unless high flow rates are used. Helium is difficult to ionize because of its higher arc resistance (higher than that of argon or carbon dioxide). This necessitates a higher voltage—which produces a much hotter arc—to support the arc. There is a noticeable increase in both the heat and temperature of a helium arc, which makes it easier to make welds on aluminum. Aluminum is highly conductive, and on thick sections, much of the welding heat can be drawn away from the weld. The use of 100% helium or helium–argon mixtures allows large welds with deep penetration to be made on thick aluminum sections.

Carbon Dioxide (CO_2)

In GMA welding of steels, 100% carbon dioxide is widely used as a shielding gas because it allows a higher welding speed, has better penetration and good mechanical properties, and costs less than the inert gases. The chief drawback in the use of carbon dioxide is the less-steady-arc characteristics and considerable weld-metal-spatter losses. The spatter can be kept at a minimum by maintaining a very short, uniform arc length. Consistently sound welds can be produced using carbon dioxide shielding, provided a filler wire with the proper deoxidizing additives is used.

Carbon dioxide is provided as a liquified gas, which means that the cylinder pressure will remain relatively constant as long as liquid remains in the cylinder. The pressure will fluctuate but more so with use and temperature changes than with the level of gas remaining. There are two ways of estimating how much gas is remaining in a CO_2 cylinder. You can feel for the lower temperature of the remaining liquid in a cylinder if it has been heavily used for some time. You can also determine the amount of gas by weighing the cylinder. A full cylinder can have as much as 50 pounds of liquid. Some people try to determine the level by tapping on the cylinder, but this is unreliable at best and can damage the finish on the cylinder. Thus tapping is not recommended as a way of determining the level of liquid in the cylinder.

GAS FLOW RATE

The shielding gas flow rate is measured in cubic feet per hour (cf/h) or in metric measure as liters per minute (l/min). The flow is metered or controlled by opening a small valve at the base of the flowmeter. The reading is taken from a fixed scale that is compared to a small ball floating on the stream of gas. Meters from various manufacturers may be read differently—from the top, center, or bottom of the ball (Figure 4-1). The ball floats on top of the stream of gas inside a tube that gradually increases in diameter in the upward direction. The increased size allows more room for the gas flow to pass by the ball. If the tube is not vertical, the reading is not accurate, but the flow is unchanged. Also, when using a line flowmeter, it is important to have the correct pressure. Changes in pressure will affect the accuracy of the flowmeter reading. To get accurate readings, be sure the gas being used is read on the

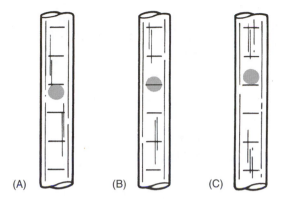

Figure 4-1 Three methods of reading a flowmeter: (A) top of ball, (B) center of ball, and (C) bottom of ball.

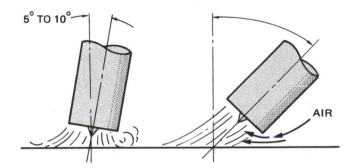

Figure 4-2 Too steep an angle between the torch and work may draw in air.

proper flow scale. Less dense gases, such as helium and hydrogen, will not support the ball on as high a column with the same flow rate as a denser gas such as argon.

The rate of flow should be as low as possible and still give adequate coverage. High gas flow rates waste shielding gases and may lead to contamination that comes from turbulence in the gas at these high flow rates. Air is drawn into the gas envelope by a venturi effect around the edge of the nozzle. Also, the air can be drawn in under the nozzle if the torch is held at too sharp an angle to the metal (Figure 4-2).

The larger the nozzle size, the higher is the flow rate permissible without causing turbulence. Larger nozzle sizes may restrict your visibility of the weld. Smaller nozzle sizes can use lower flow rates but may not provide good shielding for larger-sized welds.

The flowmeter may be merely a flow regulator used on a manifold system, or it may be a combination flow and pressure regulator used on an individual cylinder (Figure 4-3).

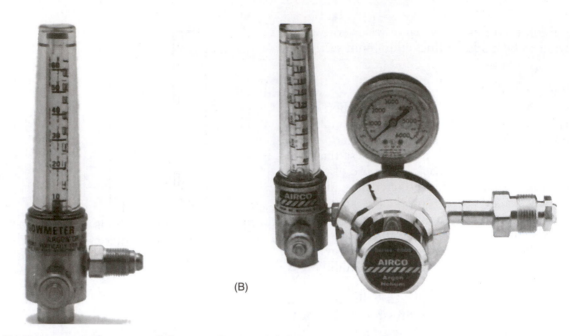

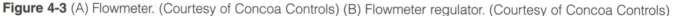

(A)

(B)

Figure 4-3 (A) Flowmeter. (Courtesy of Concoa Controls) (B) Flowmeter regulator. (Courtesy of Concoa Controls)

SHIELDING GAS AND COST

The cost of the gas is not the only factor that must be considered when selecting a shielding gas for GMA welding. For example, using CO_2 will produce the most spatter, and the average efficiency will be about 93%. Using a 75%-argon, 25%-CO_2 gas mixture will result in somewhat less spatter and an efficiency of approximately 96%. A 98%-argon, 2%-oxygen mixture will produce even less spatter, and the average efficiency will be about 98%.

Welding speeds can also be affected by the shielding gas. Both argon–oxygen and argon–CO_2 mixtures can be used for spray arc-metal transfer, whereas straight CO_2 cannot. Spray arc provides the highest metal transfer rate, and that means higher productivity. But that is a benefit only if you are welding on thick sections that can withstand the higher heat of this transfer method. In other words, there may not be a savings using the mixed gases if you are not welding on thick stock.

SUMMARY

The shielding gas or mixture of shielding gases used for making a GMA weld can greatly affect the weld in a number of ways. It can affect the weld's appearance, welding speed, welding efficiency, and other such factors. It is important to learn how each shielding gas reacts when used for welding so that you can make better decisions when selecting the gas to be used for a specific job.

REVIEW QUESTIONS

1. What are the most commonly used gases?
2. What dictates which shielding gas or mixture of gases may be used for specific welds?
3. What is argon?
4. What does adding oxygen or carbon dioxide to argon do?
5. The presence of oxygen in the shielding gas can cause what?
6. What is helium?
7. What is the benefit of using 100% helium or helium–argon mixtures?
8. What is the chief drawback in the use of carbon dioxide?
9. How is the shielding-gas flow rate measured?
10. High gas flow rates waste shielding gases and may lead to _____.
11. Using a 75%-argon, 25%-CO_2 gas mixture will result in what?

CHAPTER 5

FILLER METAL

From the days of using rivets, bolts, and nuts as fasteners, the metal industry has advanced to using lightweight welds. These welds use *electrodes,* or filler metals of similar composition, to fuse parts of structures. Changes in welding methods have resulted in stronger, safer, and faster construction at a considerable cost savings to businesses and consumers.

The wire electrode is a key component of GMAW. Its main function is to melt together with the base metal as the filler metal, forming a fusion weld (Figure 5-1).

Many types and compositions of electrodes allow the welder to select filler metals that closely match the base metal.

When welding on high-carbon-content steel, use an electrode with approximately the same carbon content. A poor electrode selection is one with a low carbon content for a base metal with high carbon properties.

GMAW ELECTRODES

The electrode used with GMAW is a solid wire. It is a carbon-steel-filler wire electrode that is continuously fed off a reel attached to the wire-feed unit. A series of drive rolls, guide tubes, a conduit liner, and a contact tube make the delivery of the electrically hot wire electrode seem effortless. When the

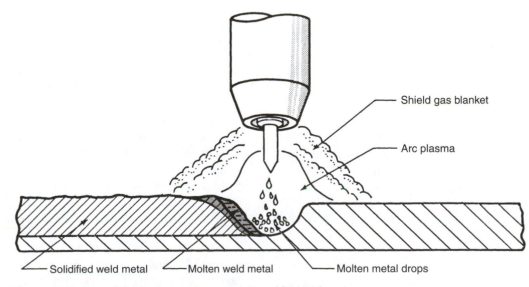

Shield gas blanket

Arc plasma

Solidified weld metal — Molten weld metal — Molten metal drops

Figure 5-1 Gas shielded metal arc welding (GMAW).

electrode contacts the base metal, an arc is established to produce the weld bead.

GMAW Electrode Specifications

AWS specifications for carbon-steel-filler electrodes used with GMAW are found in AWS publication A5.18.

When you select a filler metal, consider its metallurgical composition, which should be as close to the composition of the base metal as possible. Composition refers to several metallurgical characteristics of both the filler-wire electrode and the base metal. These metallurgical characteristics are the basic chemistry of the metal, that is, what it is made of and how it reacts to stress.

Deoxidizers, such as manganese and silicone, might be added to the electrode wire in small amounts to help stabilize the metallurgical reaction to oxygen and other undesirable gases in the base metal during solidification.

Although no flux or slag is associated with GMA welding, you will occasionally notice brownish-colored, intermittent spots on the surface of the solidified weld bead. These are impurities that have floated to the weld surface as they combined with the deoxidizers in the electrode. These impurities were in the base metal or were not removed from its surface prior to welding. The glasslike substance (surface oxides) pops off the weld bead as it cools and does not damage the weld surface.

CAUTION

Always wear your safety glasses when looking at the surface of the cooling weld bead.

Electrode Diameters

Wire electrodes are produced in 0.023″, 0.030″, 0.035″, 0.045″, and 0.062″ diameters. Other larger diameters are available for production work and can include wire diameter sizes such as 1/16″, 5/64″, and 7/64″. These wire electrodes are manufactured to meet special applications of the base metal.

Some wire electrodes may look like copper wire because a very thin copper cladding is applied as a lubricant to allow the wire electrode to slip through a tool used to manufacture the electrode called the *die*. The thin copper coating protects the electrode from small amounts of condensing moisture, retards rusting, and improves the electrical contact between the wire electrode and the contact tube.

The copper coating on the wire electrode does not affect its ability to produce sound welds. The amount of copper is so small that it either burns off or is diluted into the weld pool with no significant reaction. Manufacturers use many methods and lubricants to draw the wire through the dies, so not all wire electrodes are copper clad.

Application

The solid wire electrode is consumed as the weld travel progresses. Less smoke and weld spatter are produced because of the protective gas shield provided to the weld zone by the shielding gas, rather than by a flux-coated electrode.

When a constant voltage (CV) power source is used, the amount of spatter produced is minimized if selected current and voltage settings are within the correct operating variables for the wire electrode. Spatter is reduced because the CV power supply provides a constant burn-off rate of the electrode wire at the arc. The constant-voltage feature keeps the voltage within the acceptable recommended voltage variable of the wire electrode, allowing the welder to have more flexibility with movement of the GMAW gun tube to work distance. The CV machine is programmed to constantly provide its maximum potential. For this reason, the CV power source is sometimes referred to as a constant-potential (CP) power source.

GMAW ELECTRODE CLASSIFICATION

The American Welding Society (AWS) has a standardized method of identifying GMA welding electrodes. This standard uses a series of letters and numbers to group filler metals into specific classifications. Within a grouping of filler metals, manufacturers are allowed to make changes in specific alloys as long as the weld produced with the electrode meets the group's specifications.

The American Welding Society's classification system for GMAW filler metals starts with the prefix *ER*. The *E* indicates that the wire is classified as an electrode. The *R* indicates a length of the rod when used for another welding process. It also tells the welder that the wire can be used as a nonconductive filler rod when dipped into a liquid weld pool.

The length of the GMAW electrode is not important because the electrode packaging comes in a variety of sizes: reels, spools, or drums (Figure 5-2). The *R* may or may not be used.

(A)

(B)

(C)

Figure 5-2 FCAW filler metal weights are approximate. They will vary by alloy and manufacturer. (A, B, and C courtesy of the Lincoln Electric Company)

The *ER* is followed by a number designating the minimum tensile strength of the deposited weld metal in thousands. For example, ER70 indicates that the weld metal, as deposited, will obtain a minimum tensile strength of 70,000 pounds per square inch.

ER70 is followed by an *S*, which indicates that this electrode is a solid-wire electrode.

The last digit, as in ER70S-1, consists of a number (1, 2, 3, 4, 5, or 6) or the letter *G*. This number-and-letter system indicates the electrode's filler metal composition and the manufacturer's recommendations for the current setting and shielding gas.

ER70S-1

With ER70S-1, the 1 represents the electrode class, that is, the properties the electrode will produce with a specific shielding gas cover.

ER70S-2

With ER70S-2, the 2 indicates that the electrode is a deoxidized mild steel filler wire and can be used on metal that has a light cover of rust, or oxide. This electrode is a general-purpose electrode that can be used in flat, horizontal, vertical, and overhead welding positions or in all weld positions. The weld pool is not very fluid, making this electrode filler a good choice for out-of-position welding or when short-arc metal transfer is selected. Shielding gases such as CO_2, Ar/CO_2, and Ar/O gas mixtures may be used with this filler electrode.

ER70S-3

With ER70S-3, the 3 indicates that this electrode does not have the deoxidizers needed for welding over a metal surface that has rust on it. The weld area must be clean and brushed to a bright metallic sheen. This electrode is used for repairs on farm equipment, home appliances, and on light-gauge metals used on automobiles.

ER70S-3 can be used with either CO_2, Ar/CO_2, or Ar/O shielding gases. This electrode uses high welding currents and may result in low-strength welds. It is recommended for use on thinner steels of low-to-medium carbon content where the tensile strength will exceed the strength of the base metal. It maintains a degree of ductility in the weld. The weld pool is more fluid than the ER70S-2, but the electrode can still be used in all positions.

ER70S-4

With ER70S-4, the 4 shows that the wire electrode contains a higher level of a deoxidizer (such as silicone) than the ER70S-3. This addition results in improved soundness of welds made with this filler. Under the same conditions, the weld bead will be flatter and slightly wider than beads made with other fillers.

The ER70S-4 electrode is commonly used when welding on structural steels, pipe, in shipyards, and many in metal-fabricating job shops. This electrode also works well with CO_2, Ar/CO_2, or Ar/O shielding gases. Metal transfer is accomplished by either the short-circuit or the spray-transfer method. The ER70S-4 electrode performs well in all welding positions, but it is difficult to weld out of position when using the spray transfer method. It can be used with pulsed-arc transfer.

ER70S-5

With ER70S-5, the 5 shows that this electrode is used in the flat position only. The weld pool will be extremely fluid. Short-circuit weld metal transfer should not be used.

This filler electrode is excellent for spray-arc metal transfer with a larger-diameter wire electrode used in the flat position on heavy or thick materials. ER70S-5 has very good deoxidizing characteristics and works best with Ar/O or Ar/CO_2 shielding gases. You can weld over rusty surfaces and maintain weld quality with this electrode.

ER70S-6

With the ER70S-6, the 6 shows this is an electrode with the highest levels of manganese and silicone for strength and deoxidation. This electrode is used on thick or thin sections. It works well over rusty surfaces or areas that cannot be easily cleaned and on light-gauge metals used in the automobile industry. And, as an all-position electrode, the short-circuit method of weld metal transfer can be used.

ER70S-6 welds a smooth bead with a uniform appearance. The filler metal has lower mechanical properties. It can be used with CO_2, Ar/CO_2, or Ar/O shielding gases.

ELECTRODE CAST AND HELIX

The electrode wire is wound (rolled) on spools, reels, or coils made of formed metal, wood, pressed fiber, or plastic called the *electrode package*. Sometimes drums are used in which the wire is coiled inside the drum.

Feed out several feet of wire electrode, and snip it off. Let it fall to the floor, and observe that it forms a circle. One complete circle or the diameter across the circle is known as the *cast* of the wire.

Now snip off the same wire to have a few inches more than just a complete circle. The wire electrode does not lie flat, and one end will be slightly higher than the other. This is the *helix*, which is measured in inches, and is the distance from the flat surface to the highest point of the wire.

The manufacturer purposely produces the cast and helix as part of the manufacturing process so the wire electrode feeds smoothly off the electrode package.

The cast and helix twist of the wire cause a rubbing effect on the inside of the contact tube. The slight bend in the electrode wire ensures a positive contact surface with the energized contact tube and electrically energizes the electrode as it travels to the base metal to establish the arc for welding (Figure 5-3).

Care of Electrodes

Wire electrodes may be wrapped in sealed plastic bags to protect the electrodes from the elements. Others may be wrapped in a special paper, and some are shipped in cans or cardboard boxes.

A small paper bag of a moisture-absorbing material (crystal desiccant) is sometimes placed in the shipping containers. This material protects the wire electrodes from rust-causing moisture.

Some wire electrodes require storage in an electric-rod oven to prevent contamination from excessive moisture. Read the manufacturer's recommendations located in or on the electrode shipping container for information on use and storage.

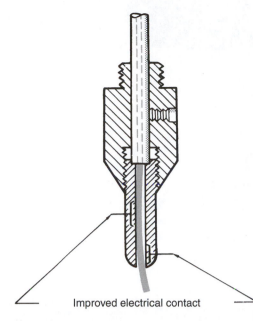

Improved electrical contact

Figure 5-3 Cast forces the wire to make better electrical contact with the tube.

The electrode may develop restrictions due to the tangling of the wire, or it may become oxidized with excessive rusting if the wire electrode package is mishandled, thrown, dropped, or stored in a damp location. Always keep the wire electrode dry, and handle it as you would any important tool.

SUMMARY

Not all GMA-welding filler metal that is classified under the same AWS identification number may be exactly the same. For example, not all ER70S-1 electrodes are exactly the same. When selecting a filler metal, you may want to test several different wires from more than one manufacturer.

Proper handling and storage of the electrodes is also important to ensure consistent welding quality.

REVIEW QUESTIONS

1. What is the main function of the wire electrode?
2. The electrode used with GMAW is a _____.
3. What should be considered when selecting a filler metal?
4. What are metallurgical characteristics?
5. What diameters are wire electrodes produced in?
6. Why is the CV power source sometimes referred to as a constant potential?
7. What is the AWS standardized method of identifying GMA welding electrodes?
8. With the ER70S-3, what does the 3 indicate?
9. When is the ER70S-4 commonly used?
10. With the ER70S-6, what does the 6 show?
11. What is one complete circle or the diameter across the circle known as?

CHAPTER 6

WELD METAL TRANSFER METHODS

The mode of metal transfer is the mechanism by which the molten weld metal is transferred across the arc to the base metal. Modes may be short-circuit transfer (GMAW-S), spray transfer, globular transfer, and pulsed-arc transfer (GMAW-P).

The chosen mode of metal transfer depends on the welding power source, the wire-electrode size, type and thickness of material, type of shielding gas used, and the best welding position for the task.

SHORT-CIRCUIT TRANSFER—GMAW-S

As the gun trigger is depressed, the wire-feed unit is energized. The wire electrode advances and bridges the distance from the contact tube to the base metal, making contact with the base metal and completing a circuit. The electrode shorts and becomes so hot that the base metal melts and forms a weld pool. The electrode melts into the weld pool and burns back toward the contact tube, extinguishing the arc, and forming a droplet or a small ball. At that moment a transfer of metal has taken place. The short-circuit process is repeated approximately 20 to 200 times per second (Figure 6-1).

The short-circuit mode of transfer is the most common method used with GMAW:

- on thin or properly prepared thick sections of material.
- on a combination of thick to thin materials.
- with a wide range of electrode diameters.
- with a wide range of shielding gases.

The 0.023″, 0.030″, 0.035″, and 0.045″ wire electrodes are recommended for the short-circuiting mode. The shielding gas used on carbon steel is carbon dioxide (CO_2) or a combination of 25% CO_2 and 75% argon (Ar).

The recommended weld position is the position in which the workpiece is placed for welding. All welding positions can be used for short-circuit transfer, but we will concentrate on the flat and horizontal welding positions for now. In the flat welding position, the workpiece is placed flat on the work surface. In the horizontal welding position, the workpiece is positioned perpendicular to the work surface.

The amperage range may be from as low as 35 for materials of 24 gauge to amperages of 225 for materials up to 1/8″ thick on square-groove weld joints. Thicker base metals can be welded if their edges are properly prepared and cut at an angle (beveled) to accept a complete joint weld penetration.

Short-circuit transfer is the most widely used of the metal transfers for general repairs to mild steel.

GLOBULAR TRANSFER

The globular transfer process is rarely used by itself because it transfers the molten metal across the arc in much larger droplets. It is used in combination with pulsed-spray transfer.

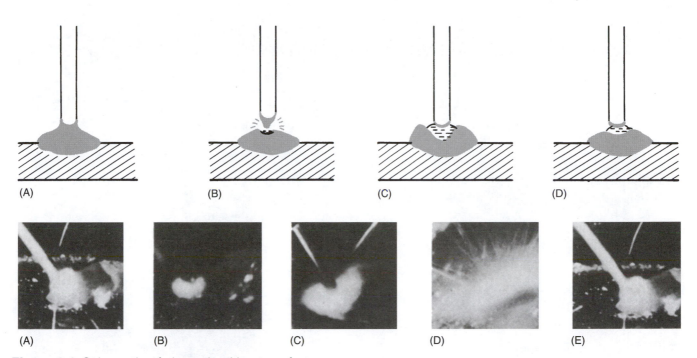

(A)   (B)   (C)   (D)

(A)   (B)   (C)   (D)   (E)

Figure 6-1 *Schematic of short-circuiting transfer.*

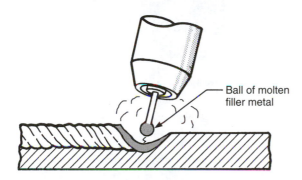

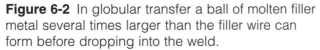

Figure 6-2 *In globular transfer a ball of molten filler metal several times larger than the filler wire can form before dropping into the weld.*

Globular transfer is generally used on thin materials and at a very low current range. It can be used with higher current but is not as effective as other welding modes of metal transfer (Figure 6-2).

SPRAY TRANSFER

Spray transfer is a popular process used in manufacturing where high deposition rates are required and deep penetration is desired. It uses a mixture of 95-98% argon and 2-5% oxygen. The added percentage of oxygen allows greater weld penetration.

This process generally uses larger-diameter wire electrodes, so it requires a higher amperage range. The higher the amperage range, the faster the weld

bead progresses and the groove joint filled. Increased current flow combines with the high percentage of argon, causing a pinching effect on the molten ball of wire electrode much like the effect pinching has on a rubber water hose. When you pinch the end of the rubber water hose, the water exits the hose in a spray pattern.

Spray-transfer process is very hot and virtually free of spatter. The sound produced by spray transfer is a quiet, hissing sound, unlike the short-circuit process that makes a raspy, frying sound. A disadvantage of spray transfer is that it produces a very fluid weld pool and is thus limited to flat and horizontal positions. It is UV intense, so you need extra burn protection for your eyes, hands, and arms (Figure 6-3).

CAUTION

A darker filter lens is required when using the spray-transfer mode.

PULSED-ARC TRANSFER—GMAW-P

The current produced by the pulsed-arc transfer mode (GMAW-P) is a dual-pulsed current. One pulse of current is a spray-transfer mode. The other pulse of current is at the higher end of the range of current provided by the globular-transfer mode.

(C)

(A)

(B)

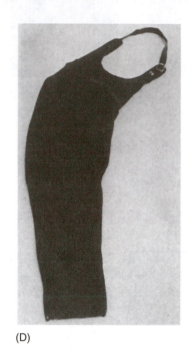

(D)

Figure 6-3 (A) All-leather, gauntlet-type, welding gloves. (B) Full leather jacket. (Courtesy Elliott Glove) (C) For welding that requires a great deal of manual dexterity, welders can wear soft leather gloves. (D) Full leather sleeve.

Pulsed-arc transfer permits the use of the spray-transfer mode at a much lower current level than commonly used. This allows the spray release at evenly spaced pulses, not continuously. Pulsed current will peak out at the current range in the spray mode and bottom out at the current range in the globular-transfer mode.

Pulsed-arc transfer permits the use of sprayed electrode wire at a low current level, so GMAW can be performed on lighter-gauge base metals. It allows welding in other than the flat position and produces little or no spatter.

The advantage of the pulse-transfer mode is that the two levels of pulse current provide the depth of penetration offered by the spray current and the ability to control the weld pool when welding thick materials to thin ones or when welding out of position. In welder's jargon these welds are often known

Figure 6-4 GMA pulsed-arc welding system. (Courtesy of CRC-Evans Automatic Welding)

as out-of-position welding, meaning other than the flat position. The disadvantage of pulsed spray is that the cost of equipment makes it more expensive than other processes (Figure 6-4).

GMA WELDING VARIABLES

Welding variables refer to adjustments or corrections made to the equipment before or during the weld operation.

You must preselect the following welding variables before starting to weld:

- The type of welding wire
- The type of shielding gas
- The desired rate of gas flow
- The size of the wire electrode

These preselected variables are dependent on the following factors:

A) The type of metal

B) The thickness of the metal

C) The position of the materials being welded

Also select the following primary adjustable variables:

D) The arc voltage

E) The speed of travel

F) The current

These primary adjustable variables control the depth of penetration, the width of the deposited weld bead, and the rate of weld metal deposited.

The following additional variables can cause changes in the primary variables:

G) Tube-to-work distance

H) Speed of the wire electrode

I) Gun angle

J) Environmental conditions

If any one of these variables is changed, a desired or an undesired change in the weld bead formation may occur. Variables A through F (except variable E) are either selected before welding starts or are settings recommended by electrode manufacturers and are not easily changed or adjusted while welding. Variables G through J can be easily changed or adjusted by the welder and can be either beneficial or detrimental to the weld pool.

Electrode Extension

The tube-to-work distance is known as the electrode extension or the length of electrode extending beyond the end of the contact tube. It refers to how far away you hold the contact tube from the base metal as you work. The size of the wire electrode will determine the distance of the tube to work.

The wire-electrode classification (numbering) system established by AWS enables you to select the correct process, electrode, and shielding gas for your task. The length of the electrode extension for a given rate of wire-feed speed has an effect on the weld pool. An excessive length of electrode extension causes resistance in the electrode. This resistance causes the electrode wire to become too hot, removing heat from the arc. Poor depth of penetration results.

CAUTION

Hold the electrode extension as close to the weld pool as possible or within the manufacturer's recommendations without dipping the contact tube into the molten weld pool. Practice is required. The arc will lose less heat and maintain maximum heat input to the weld pool. Weld penetration into the base metal will be improved.

Electrode extension and tube to work are terms you have read about. The cut-away drawing in Figure 6-5 shows the contact tube recessed into the gun

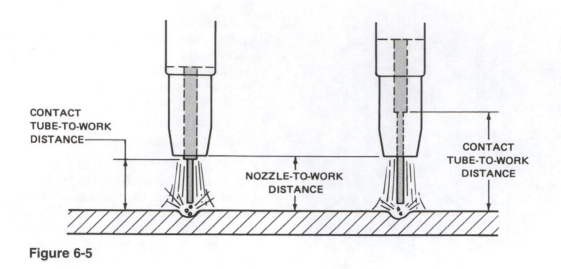

Figure 6-5

nozzle. Some operations require an extended contact tube or that it be flush with the nozzle. We recommend that you use the recessed tube. By recessing the tube 1/8″ into the nozzle, you stand less chance of the contact tube's being dipped into the weld pool. To stay in the given variable, you must be aware of the contact tube's location in relation to the nozzle.

SPEED OF THE WIRE ELECTRODE

The wire-feed speed is generally recommended by the electrode manufacturer and is selected in inches-per-minute (ipm) (this number indicates how fast the wire exits the contact tube). The welder uses a wire-speed control dial on the wire-feed unit to control the ipm rate. It can be advanced or slowed to control the burn-off rate, which is how fast the electrode transfers into the weld pool, to meet the welder's skill in controlling the weld pool (Table 6-1).

To accurately measure wire-feed "ipm," snip off the wire at the contact tube. Squeeze the trigger for 15 seconds, release it, and snip off the wire electrode. Measure it and multiply its total length by four. The result is how much wire is fed in inches per minute.

GUN ANGLE

The work angle is the position of the welding gun in relation to the joint configuration of the base metal. The gun position is described with two distinctive names, transverse and longitudinal angles. The gun angle, or work angle, is the angle at which you position the gun to the base metal (Figure 6-6).

TABLE 6-1

TYPICAL AMPERAGES FOR CARBON STEEL				
WIRE FEED SPEED* IN/MIN (M/MIN)	WIRE DIAMETER AMPERAGES			
	.030 IN (0.8 MM)	.035 IN (0.9 MM)	.045 IN (1.2 MM)	.062 IN (1.6 MM)
100 (2.5)	40	65	120	190
200 (5.0)	80	120	200	330
300 (7.6)	130	170	260	425
400 (10.2)	160	210	320	490
500 (12.7)	180	245	365	—
600 (15.2)	200	265	400	—
700 (17.8)	215	280	430	—

*To check feed speed, run out wire for 1 minute and then measure its length.

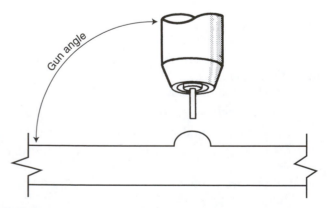

Figure 6-6 Gun angle.

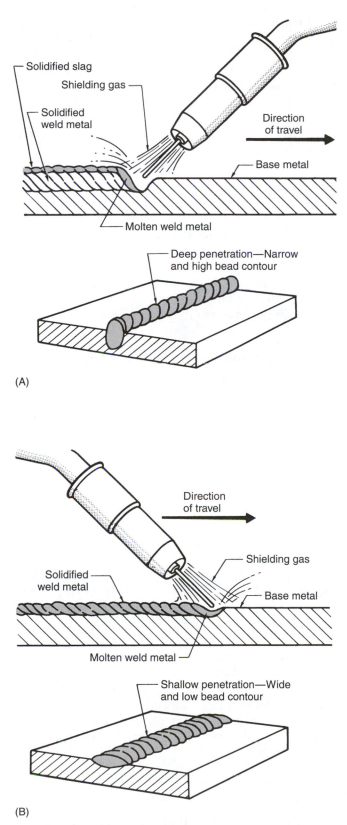

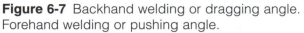

Figure 6-7 Backhand welding or dragging angle. Forehand welding or pushing angle.

Travel Angle

The *longitudinal* angle, or travel angle, is measured in the direction you will travel as the weld bead progresses. It is the angle between the center line of the welding gun position and a line perpendicular to the axis of the weld bead.

Forehand and Backhand Techniques

Both the longitudinal and the transverse angles remain the same regardless of whether you are pushing or pulling the weld bead. Pushing a weld bead is the *forehand* technique, and pulling, or dragging, is the *backhand* technique (Figure 6-7).

Advantages of the Backhand Technique

By using the backhand technique, you can readily see the bead as it progresses, and the amount of spatter is minimized. The gun position is easier to maintain, ensuring a consistent depth of penetration. The backhand method seems the more natural way to weld.

This method may result in a slightly slower weld progression due to the amount of weld reinforcement required. The slower-moving backhand method allows a slower progression of the weld. This longer welding time at the base metal provides a preheat time in the weld zone as the weld bead progresses. The preheat time allows the weld bead to flow easily into the weld pool, keeping spatter to a minimum.

Disadvantages of the Backhand Technique

The weld bead will have a more pronounced weld face when you use the backhand technique. Because of the convex, or raised or rounded, shape of the weld bead, more work is required if the product has to be finished by grinding with a power tool. The surface of the weld bead is ground flush, or blended into the surface of the base metal so the welded joint is not noticeable.

When you use the backhand technique, the weld joint is harder to follow because of the welder's hand position and the GMAW gun position. Because of these obstructions, it is easier to wander from the seam of the weld joint. An inexperienced welder sometimes directs the wire too far back into the weld pool, causing the wire to build up in the face of the weld pool. Loss of depth of penetration, or depth that fusion extends into the base metal during GMAW, occurs. Weld reinforcement will be excessive, which increases welding time and wastes welding electrode material.

Advantages of the Forehand Technique

An advantage of the forehand technique is that you can readily see the joint where the bead will be deposited. The contact tube is easier to see, ensuring a more consistent wire electrode extension. Weld progression is slightly faster with this method. Less weld reinforcement is applied to the weld joint at the leading edge of the weld pool, where you are able to see new metal. This is metal that has not yet been welded but is in the direct path of the weld pool as the weld progresses, eliminating the weld pool buildup that many novice welders encounter. Contact with the new metal is improved, resulting in good depth of penetration and less weld face buildup. The weld bead will not be as pronounced at the weld face and saves work if the finished product is to be ground flush.

Disadvantages of the Forehand Technique

The amount of spatter is slightly increased with the forehand technique if a 10°-15° longitudinal angle is not maintained, and the hot weld pool moves into the unpreheated new metal (base metal). This may cause a rougher surface on the weld bead face, and the weld bead ripple may not be as smooth as one produced using the backhand method.

Depth of penetration is not as great with the forehand method as it is with the backhand method. The weld progresses so quickly that the arc does not preheat the weld zone as well as the slower-moving backhand method does. Discuss the pros and cons of the two welding techniques with your instructor.

TRAVEL SPEED

AWS defines *travel speed* as the linear rate at which the arc is moved along the weld joint. Fast travel speeds deposit less filler metal. If the rate of travel increases, the filler metal cannot be deposited fast enough to adequately fill the path melted by the arc. This causes the weld bead to have a groove melted into the base metal next to the weld and left unfilled by the weld. This condition is known as *undercut*. Its location along the edges of the weld bead is called the *toes* of the weld bead. Slower travel speeds will, at first, increase penetration and increase the filler weld metal deposited. As the filler metal increases, the weld bead will build up in the weld pool. Weld penetration is decreased if the travel speed does not contact new metal and stay on the leading edge of the weld pool.

If all welding conditions are correct and remain constant, the preferred rate of travel for maximum weld penetration is a travel speed that allows you to stay within the selected welding variables and still control the fluidity of the weld pool. This is an intermediate travel speed, or progression that is neither too fast nor too slow.

Another way to determine correct travel speed is to consult the manufacturer's recommendations chart for the ipm burn-off rate for the selected electrode.

TUBE TO WORK

AWS refers to *tube to work* as the electrode extension, or the length of electrode extending beyond the end of the contact tube. Most welders use the jargon *tip to work, stick out,* or the approved AWS term, *tube to work*. Electrode extension or tube to work refer to how far away you hold the contact tube from the base metal as you work.

ENVIRONMENTAL CONDITIONS

Weather conditions affect the welder's ability to make quality welds. Humidity increases the chance of moisture entering the weld zone. Water (H_2O), which consists of two parts hydrogen and one part oxygen, separates in the weld pool. Hydrogen is trapped and causes an undesirable result, called *hydrogen entrapment,* a situation in which only one part of hydrogen is expelled. Hydrogen entrapment can cause weld beads to crack or become brittle. The evaporating moisture will also cause a series of small holes, called *porosity,* to form in and through the weld bead.

If the weld area is subjected to wind or airflow greater than what the envelope of shielding gas can tolerate, the protective atmosphere is removed, causing atmospheric contamination and porosity.

To prevent hydrogen entrapment, porosity, and atmospheric contamination, you may find it necessary to preheat the base metal to drive out moisture. Storing the wire electrode in a dry location is recommended. Work indoors away from drafts or windy areas. Install barriers to prevent these conditions, and increase the shielding gas flow to the maximum recommended for the task.

SUMMARY

The type of metal transfer, welding gun angle, electrode extension, and so on are often specified in a welding procedure. In those cases the welder must use the information that is provided. Most day-today welding is not done under such strict guidelines. For this reason you must practice using all of the different metal-transfer methods. Having learned all of these methods in school will aid in your being a better welder in the field.

REVIEW QUESTIONS

1. What is the most widely used metal transfer for general repairs to mild steel?
2. Globular transfer is generally used on what?
3. What is the disadvantage of spray transfer?
4. What is the advantage of the pulse-transfer mode?
5. What is the disadvantage of pulsed spray?
6. What determines the distance of the tube to work?
7. What is the transverse angle?
8. What is the longitudinal angle?
9. What is the advantage of the forehand technique?
10. What is the AWS definition of travel speed?
11. What is porosity?
12. What is necessary to prevent hydrogen entrapment, porosity, and atmospheric contamination?

CHAPTER 7

EQUIPMENT SETUP

Although each manufacturer's GMA welding equipment is designed differently, they are all set up in a similar manner. It is always best to follow the specific manufacturer's recommendations regarding setup as provided in their equipment literature. You will find, however, that, in the field, manufacturer's literature is not always available for the equipment you are asked to use. It is therefore important to have a good general knowledge and understanding of the setup procedure for GMA welding equipment. Figure 7-1 shows the various components that compose a GMA welding station.

POWER SUPPLY

As was mentioned in Chapter 3, the welding power supply can be either a transformer, rectifier, or engine-driven generator. The welding portion of both of these types of equipment requires similar setup procedures. For specifics regarding the setup, care, and maintenance for the engine portion of an engine-driven welder, you must refer to the materials supplied by the manufacturer because improper care can result in serious damage to the engine.

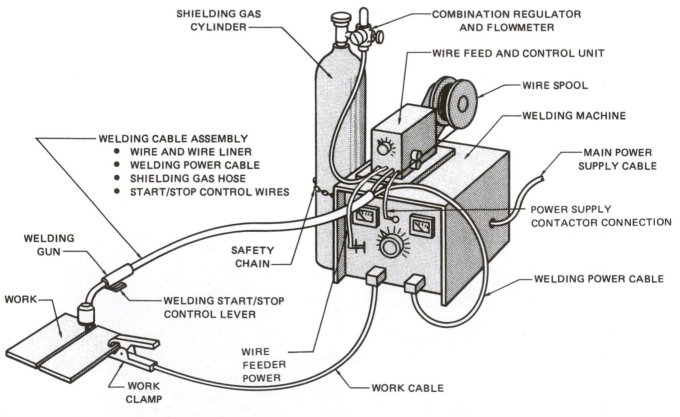

Figure 7-1 Gas metal arc welding equipment.

Welder Connections

Many industrial-type GMA welding machines and wire-feed units come as separate devices (Figure 7-2). These units require some assembly before the welder can be used. Many smaller GMA welding units and some larger commercial units have the wire-feed mechanism built into the welder, so no assembly is required (Figure 7-3).

There are a number of advantages in having the wire-feed assembly separate from the welding machine, including the following:

- Portability–The wire-feed unit can be moved some distance from the welding machine. This will allow you to use shorter welding-gun assembly cable, thus reducing wire-feed problems, while still allowing greater access to welds. Unlike the limitations on the length of the gun assembly, there is virtually no limitation to the interconnecting power and control cables between the welding machine and wire-feed units.

- Flexibility–Having the wire-feed unit separate will allow you to change the wire-feed assembly without the necessity of purchasing another complete welding machine. This makes it easier to use different types of wire-feed units for special applications. Wire-feed units such as a standard wire-feed system, spool gun, and machine-welding assembly can all be easily interchanged with the same welding power supply (Figure 7-4).

- Space savings–Welding machines can be located out of the way (for example, under a welding table, some distance from the welding area, or even overhead), and the wire-feed unit can be suspended at the end of a specially designed

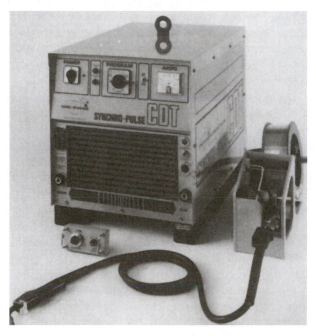

Figure 7-2 GMA pulsed-arc welding system. (Courtesy of CRC-Evans Automatic Welding)

Figure 7-3 A 90-ampere power supply and wire feeder for welding sheet steel with carbon dioxide shielding. (Courtesy of Miller Electric Mfg. Co.)

Figure 7-4 Modern wire feeder with digital preset and readout of wire-feed speed and closed-loop control. (Courtesy of Miller Electric Mfg. Co.)

boom. This keeps both the welder and the wire-feed unit out of the way of the operation and clear of sparks and hot metal (Figure 7-5).

- Equipment replacement–With separate components, it is easier to upgrade either the welding power supply or wire-feed unit when such changes become necessary.

Figure 7-5 GMA spot-welding machine. (Courtesy of Miller Electric Mfg. Co.)

Setting Up the Wire-Feed Unit

There are several interconnecting cables between the wire-feed unit and the welder on systems that have a separate wire-feed assembly. When the manufacturer's assembly instructions are available, follow them. But if they are not available, the following generic instructions can be used. Most GMA welding uses DC electrode-positive current. This means that one end of the power cable will be connected to the welding power source's positive terminal, and the other end of the cable will be connected to the welding power terminal on the wire-feed unit. The welding power terminal of the wire feed unit would be located on or near the end of the welding gun assembly.

The welding gun assembly usually uses a bayonet-type slip end connector. Each welding gun manufacturer and welding machine manufacturer uses specially designed connectors. Adapters are available that will allow you to connect different manufacturer's guns and wire-feed units together (Figure 7-6). Often there are shielding gas O-rings on the slip connector of the welding gun assembly. These O-rings usually do not require lubrication for assembly or to form a gas-tight seal. However, if assembly is difficult or a gas-tight seal is not formed, a small quantity of a silicone-based lubricant can be used.

The gun assembly connector is usually held in place with either a spring clip or set screw. The spring clip fits through a hole in the housing and passes through a notch in the connector.

A control cable connects the wire-feed unit to the welding power source. This cable is used to stop and start the welding current each time the gun trigger is depressed. This allows the welding power to be turned on and off remotely by the welder.

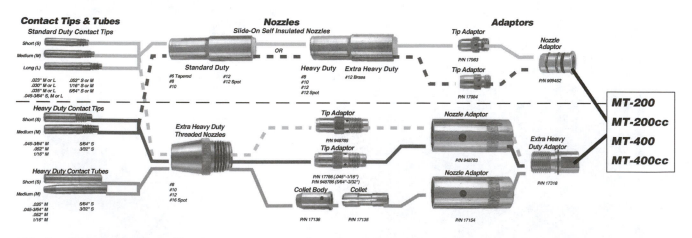

* Using Heavy Duty contact tips or tubes and Extra Heavy Duty threaded nozzles will increase the rated current capacity of the MT-400 or MT-400cc to 500 amps.

———— Primary use of accessories. ▪▪▪▪▪▪ Alternate use of accessories.

Figure 7-6 Accessories and parts selection guide for a GMA welding gun. (Courtesy of ESAB Welding and Cutting Products)

The wire-feed power may be provided by plugging it into the welding machine or into a weld receptacle. In most cases the wire-feed unit is powered with 110 volts (regular household current). In most cases wire feed units require a circuit capable of providing at least 20 amps of electrical power.

Shielding gas is provided by a hose connected between the wire-feed unit and either a gas cylinder or manifold system. The hose that connects the shielding gas to the wire-feed unit must be of an approved type that will not contaminate the shielding gas. Do not use a hose that has been used for any other purpose such as for water, hydraulic fluid, or fuel gases because these materials can leave residues in the hose that can cause contaminations in the weld. In most cases the shielding gas hose is connected with a threaded connector. Be sure the fittings are tightened and leak-checked before you begin welding.

The work clamp sometimes referred to as the ground clamp cable is attached to the welding power supply's negative terminal. This terminal may be identified by the minus symbol or the words *ground* or *work* (all three are often used interchangeably). The work lead must be of an adequate size to handle the welding current. Table 7-1 lists wire gauge sizes, welding currents, and links that can be used to determine the correct size cable required for your particular application.

TABLE 7-1

COPPER AND ALUMINUM WELDING LEAD SIZES

COPPER WELDING LEAD SIZES

AMPERES		100	150	200	250	300	350	400	450	500
ft	m									
50	15	2	2	2	2	1	1/0	1/0	2/0	2/0
75	23	2	2	1	1/0	2/0	2/0	3/0	3/0	4/0
100	30	2	1	1/10	2/0	3/0	4/0	4/0		
125	38	2	1/0	2/0	3/0	4/0				
150	46	1	2/0	3/0	4/0					
175	53	1/0	3/0	4/0						
200	61	1/0	3/0	4/0						
250	76	2/0	4/0							
300	91	3/0								
350	107	3/0								
400	122	4/0								

(LENGTH OF CABLE)

ALUMINUM WELDING LEAD SIZES

AMPERES		100	150	200	250	300	350	400	450	500
ft	m									
50	15	2	2	1/0	2/0	2/0	3/0	4/0		
75	23	2	1/0	2/0	3/0	4/0				
100	30	1/0	2/0	4/0						
125	38	2/0	3/0							
150	46	2/0	3/0							
175	53	3/0								
200	61	4/0								
225	69	4/0								

(LENGTH OF CABLE)

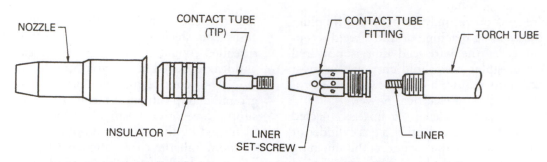

Figure 7-7 Typical replaceable parts of a GMA welding gun. (Courtesy of Hobart Brothers Company, Troy, Ohio)

Some large-amperage welding setups require cooling water to prevent the welding gun from overheating. In most cases today that welding water is provided by a recirculating system as a way of reducing the quantity of wasted water. Because of environmental concerns, many local and state governments have restricted or prohibited the use of waste water or nonrecirculating systems. If you have questions about your local laws, you can contact your city, county, or state government agencies.

When welding gun cooling water is used, supply and return water connections are provided on the wire-feed unit. These fittings are left-hand fitting to prevent the accidental connection of the cooling water to the shielding gas system. These fittings must be securely tightened to prevent water leakage. Any water leaking around electrical fittings would be extremely hazardous.

WELDING GUN AND CABLE ASSEMBLY

The welding cable assembly consists of a number of individual wires and hoses usually contained within one oil- and heat-resistant outer covering. Wires found inside the covering include the welding power cable and the control wires. There is a shielding gas hose and, if the torch is water cooled, two water hoses: one for supply and one for return water. There is also an electrode guide or liner. This liner extends the entire length of the cable assembly, and through its center the welding wire is feed to the gun end.

The welding gun consists of a number of components, some of which are considered to be consumable. The consumable parts are designed to be replaced as they wear out or become damaged during welding. Most of the other parts on the welding gun can also be replaced although they are not considered to be consumable. The welding gun handle is made of a hard heat-resistant plastic. It is form fitted to make it easier and more comfortable for a welder to hold.

The gun trigger is an on/off switch. The lever on this switch is designed to be large enough so that it is easily worked with the gloved hand.

The conductor tube fits into the gun handle. The connector tube is hollow so that the electrode liner can pass through it as well as the shielding gas. The conductor tube connects to the welding power cable in the handle. It conducts the welding current to the end of the gun assembly; therefore it is heavily insulated to prevent accidental arcing.

The end of the conductor tube is threaded both internally and externally. The external thread allows an insulating ring to be screwed on. This ring may or may not be part of the nozzle itself. If the ring is separate, it must be screwed on correctly so that the nozzle will slip over it to the proper depth. On some guns the insulator ring will be screwed on backward, but this prevents the nozzle from being installed correctly (Figure 7-7). Outside of the insulator there are metal snap rings that allow the nozzle to be held in place by friction. Conductor tubes are available as straight, 45°, 90°, and/or flexible (Figure 7-8). The most commonly used conductor tubes have a 45° angle.

The inside thread on the conductor tip is used to screw in a gas diffuser, which serves several purposes. It holds the electrode contact tube in alignment with the electrode liner, which is held in place on the diffuser by tightening a small set screw (Figure 7-9). If this set screw is not tight, the liner can pull away from the contact tube and result in the electrode wire's becoming frequently jammed. The gas diffuser also has a number of holes drilled into its side. These holes allow the shielding gas to be uniformly distributed, diffused, inside the nozzle. The contact tube screws into the end of a gas diffuser. This contact tube is the point at which welding current is transferred to the welding wire as it passes through. The

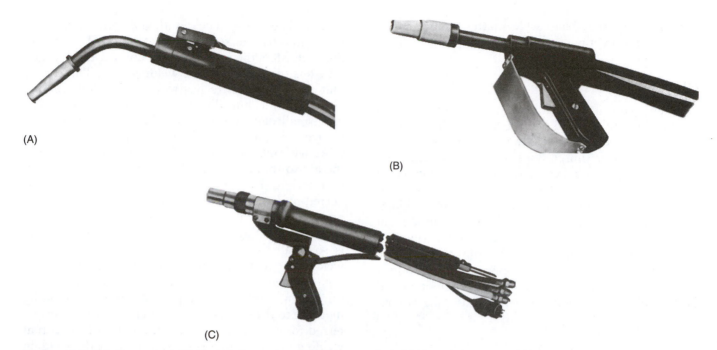

Figure 7-8 (A) Air-cooled GMA welding gun; (B) Water-cooled GMA welding gun; (C) Water-cooled GMA welding nozzle. (Courtesy of Miller Electric Mfg. Co.)

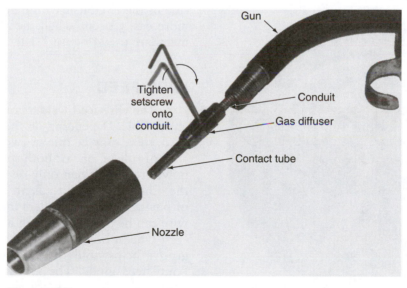

Figure 7-9

contact tube is considered to be a consumable item. Because a current is transferred to the filler wire passing through it, the precision machine hole inside the contact tube will become worn. As this wear occurs, over time the wire will not feed as smoothly as it should. The contact tube is also subject to both heat and sparks from the weld, which also over time will damage it.

The nozzle is the hollow metal tube located at the end of the gun assembly through which the welding wire and shielding gas flow into the part. Most nozzles are made of copper, although other materials such as brass and chrome plating are also used. Copper nozzles are the most popular because they resist welding spatter and can withstand the welding heat. Chrome-plated nozzles are significantly more expensive, but they resist weld spatter buildup much better than do copper nozzles.

CURRENT PATH

The path of the welding current from the welding machine to the arc and back through the work cable to the welding machine is as follows. Current leaves the power source's positive terminal and then is carried through a heavy, flexible copper cable to the power terminal on the wire-feed unit. This terminal is located on or close to the end of the welding-gun assembly. The current passes throughout the gun assembly connector and is transferred through a stranded copper cable inside the gun cable. In the welding handle the stranded copper cable is clamped onto one end of a connector block. On the other end of the connector block, the conductor tube is con-

Figure 7-10 Covered wire reel to keep wire clean and prevent accidental contact. (Courtesy of the Lincoln Electric Company)

nected. The current passes through the conductor tube and is transferred to the gas diffuser. The gas diffuser transfers the current to the contact tube, which transfers the current to the welding wire. This means that the welding wire has to carry the full welding current only a short distance to the arc. Because of its very small diameter, it would be impossible for the electrode to carry the full welding current for any distance without overheating. The current travels across the arc to the work, and the work conducts the current back to the work clamp. The work clamp is connected to a welding cable that returns the current to the negative terminal on the weld.

The welding filler metal used in GMA welding is so small that it is incapable of carrying the full welding current for more than a fraction of an inch. For this reason the welding current is transferred through a number of components from the welding machine to the gun assembly, where it is finally transferred to the filler metal near the point of the actual welding arc. It is important to note that the welding current travels back through the filler metal to the spool on the wire-feed unit. This means that if the wire-feed coil is accidentally connected to the work side of the circuit, it would arc out. Such arcing can more easily occur when metal-framed coils of filler metal are used (Figure 7-10).

WIRE FEED

The wire-feed welders on the wire-feed unit are changeable. Depending on the type of wire-feed unit used, there may be one or two sets of rollers. Within a set of rollers, one or both rollers may be "drive" or power rollers. When only one roller is a drive roller, the other is referred to as an "idler". The advantage of having two drive rollers in a roller set is that the wire can be pushed through a longer gun cable.

One or both of the rollers will be grooved. This groove helps hold the wire in alignment with the "out" feed guide. The groove in the roller may be V-shaped, U-shaped, or knurled (Figure 7-11). V-shaped rollers are most commonly used for steel, stainless steel, and other hard filler metal. U-shaped rollers are most commonly used for softer filler metals such as aluminum. Knurled rollers are used mainly on larger-diameter hard filler metals where additional roller traction is needed to provide a smooth filler metal feed.

The groove in the feed roller must be properly sized to fit the filler metal diameter being fed. The size of the filler metal to be used with a roller is usually stamped on the side of the roller (Figure 7-12). If

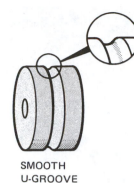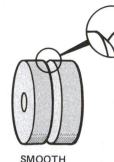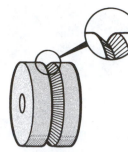

SMOOTH
U-GROOVE

SMOOTH
V-GROOVE

KNURLED
V-GROOVE

Figure 7-11 Feed rollers.

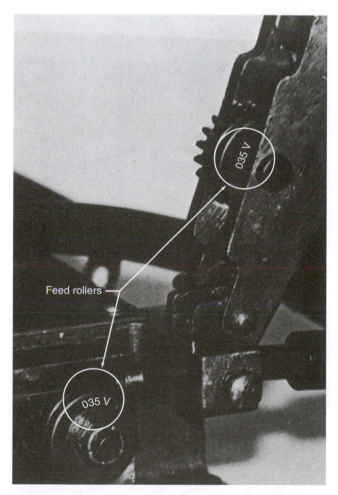

Feed rollers

035 V

035 V

Figure 7-12 Check to be certain that the feed rollers are the correct size for the wire being used.

the groove on the roller is too small for the wire being fed, the wire may wander out of the groove and may not be fed in proper alignment to the out-feed guide. If the groove in the roller is too large, the wire may not make firm enough contact with the roller to be fed (Figure 7-13).

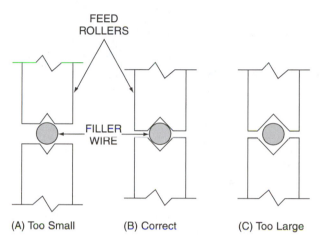

FEED
ROLLERS

FILLER
WIRE

(A) Too Small

(B) Correct

(C) Too Large

Figure 7-13 Select the correct rollers to fit the filler wire.

Out-Feed Guide

The out-feed guide should be aligned as close as possible to the feed roller and in a straight line with the groove. The farther away the guide is from the feed roller, the more likely a "bird's nest" or other misfeed problems Figure 7-14 will occur. Misalignment of the out-feed guide increases drag on the filler metal, which can lead to erratic wire feeding, and may result in scraping away small slivers of softer filler metal. Such slivers are then carried into the wire liner, where they can accumulate, resulting in wire-feed problems that can be remedied only by replacing the liner itself.

Feed-Roller Tension

In order for the feed rollers to properly grip the filler metal, they must have proper tension applied. This tension can usually be adjusted by turning a screw, bolt, or hand wheel (Figure 7-15). Tension on the feed roller that is too loose or too tight can result in wire-feed problems. If the tension is too light, the

Figure 7-14 "Bird's nest" in the filler wire at the feed rollers.

(A)

(B)

Figure 7-15 Push the wire through the guides by hand. Adjust the wire-feed tensioner.

wire feed can be erratic and may actually stop momentarily as the welder manipulates the welding gun. Too tight a wire-feed pressure can result in bird's nesting of the filler wire if the wire jams at the tip or in the liner. To set the proper adjustment, pull the gun trigger and allow the wire to feed. Be careful not to allow the wire to contact the work or other grounded metal objects. As the wire is being fed, increase the roller tension until the wire feeds smoothly. To be sure the wire is not tensioned too tightly, gently apply pressure to the wire spool with your hand. The wire-feed roller is tensioned properly when the wire spool can be stopped without excessive pressure (Figure 7-14). A secondary advantage of not having excessive wire-feed roller pressure is that if a welding cable, shielding gas hose, or other such item becomes tangled in the filler metal spool, the spool can be stopped before damage occurs to the wire, hose, or other such object.

Reel Tension

The spindle that the reeler or coil is mounted on has a friction adjustment that allows the wire spool to be stopped quickly once the welding trigger is released. Spool coasting after the trigger is released can result in two potential problems. One problem is that the loose wire can become tangled as welding is resumed. Another problem is that, when welding is resumed and slack feed is out of the wire, a momen-

tary stopping of the feed can occur. This momentary hesitation can result in a weld discontinuity or possible burn-back of the filler metal to the contact tube. The tension on the spindle is properly set when the spool is almost immediately stopped each time the welding gun trigger is released. Excessively high drag tension can result in wire-feed problems also. Increase or decrease the wire-feed tension while starting and stopping the feed, using the gun trigger until a smooth stop and restart without "hesitation" occurs.

WORK LEAD

GMA welding is very sensitive to changes in arc voltage. For that reason it is important to securely connect the work clamp to the work. A loose or poor connection will result in increased circuit resistance, which can significantly reduce the arc's voltage, thus affecting the weld's quality. A more significant voltage problem will occur when this resistance varies during the course of a weld. Such variations can dramatically—and adversely—affect your ability to maintain weld bead control. GMA welding is more significantly affected by changes in the arc voltage than is SMAW (stick) welding. To ensure a good work connection, remove any dirt, rust, oil, or other surface contamination at the point where the work clamp is connected to the weldment.

REVIEW QUESTIONS

1. What are the three types of welding power supplies?
2. What are the advantages of having the wire-feed assembly separate from the welding machine?
3. Most GMA welding uses _____ current.
4. The welding gun assembly uses what kind of connector?
5. Some high-amperage welding setups require _____ to prevent the welding gun from overheating.
6. What is the most commonly used conductor tube angle?
7. What is the purpose of the gas diffuser?
8. What is the most popular nozzle?
9. What is the advantage of having two drive rollers in a roller set?
10. What shape of rollers is most commonly used for softer filler metal?
11. Too tight a wire-feed pressure can result in what?

CHAPTER 8

WELDING SYMBOLS

Welding symbols on drawings show a welder exactly what welding is needed. They provide the welder with the required information to make the correct weld. Learning to interpret welding symbols enables a skilled welder to place the correct weld in the correct location. Welds that are not located correctly may result in greater welding expense, less serviceable weldments, and possibly even unsafe parts.

WELD JOINT DESIGN

A variety of factors must be considered when selecting joint design for a specific joint weldment. Each factor, considered alone, would result in a part that might not be able to be fabricated. For example, magnesium is very susceptible to postweld stress, and the U-groove works best for thick sections.

Joints must be designed to allow parts to be welded together in a manner that best distributes the stresses, reduces distortion, and provides for the greatest strengths. Forces applied to a joint are the following: tensile, compression, bending, torsion, and shear (Figure 8-1). A weldment's ability to withstand these forces depends on both joint design and weld integrity.

The basic parts of a weld joint design that can be changed include the following:

- Joint type—The type of joint is determined by the way the joint members come together (Figure 8-2).

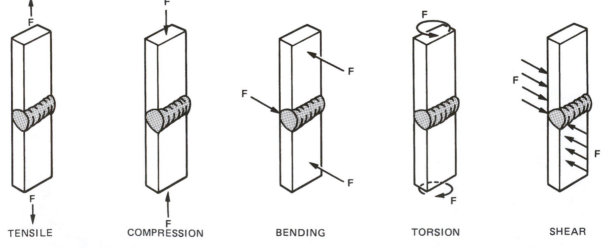

TENSILE   COMPRESSION   BENDING   TORSION   SHEAR

Figure 8-1 Forces on a weld.

- Edge preparation—The faying surfaces of the mating members that form the joint are shaped for a specific joint. This preparation may be the same on both members, or each side can be shaped differently (Figure 8-3).
- Joint dimensions—The depth and/or angle of the preparation and the joint spacing can be changed to make the weld (Figure 8-4).

PLATE WELDING POSITIONS

The ideal welding position for most joints is considered to be the flat position, which usually allows for the largest possible weld pool to be controlled. As a general rule, the larger the weld bead that can be produced in a single pass, the faster the joint can be completed.

Not all weldments can be positioned so that all of the welding can be made in the flat position. Some joints on a weldment may have to be produced in the "out-of-positions," which refers to all welds that are not produced in the flat position.

Some applications lend themselves very successfully to welds made in a position other than flat. For example, some welds made in very thin metal sections may be more easily controlled in the vertical position.

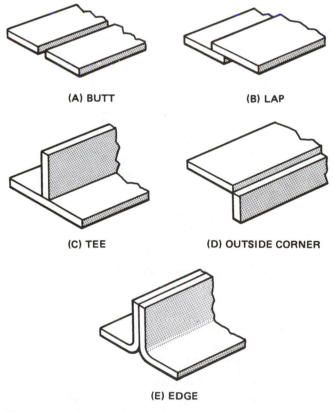

(A) BUTT   (B) LAP

(C) TEE   (D) OUTSIDE CORNER

(E) EDGE

Figure 8-2 Types of joints.

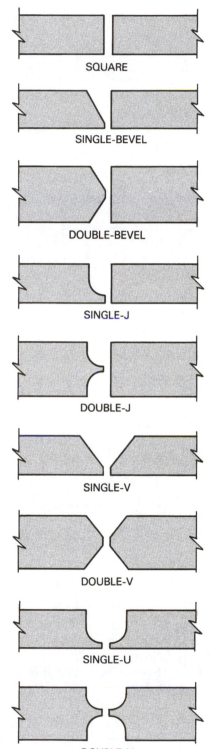

SQUARE

SINGLE-BEVEL

DOUBLE-BEVEL

SINGLE-J

DOUBLE-J

SINGLE-V

DOUBLE-V

SINGLE-U

DOUBLE-U

Figure 8-3 Edge preparation.

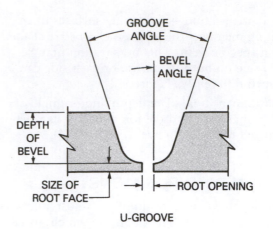

Figure 8-4

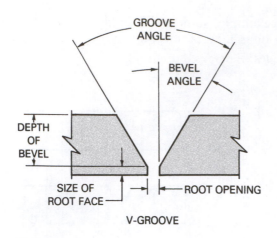

Figure 8-5 Plate flat position.

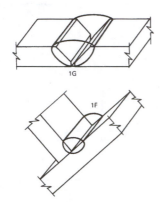

Figure 8-6 Plate horizontal position.

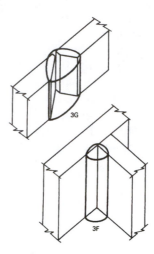

Figure 8-7 Plate vertical position.

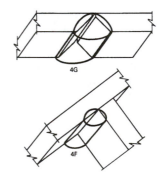

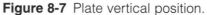

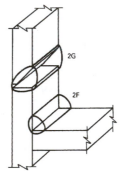

Figure 8-8 Plate overhead position.

The American Welding Society has divided plates into four basic positions for groove (G) and fillet (F) welds as follows:

- Flat 1G or 1F—Welding is performed from the upper side of the joint, and the face of the weld is approximately horizontal (Figure 8-5).
- Horizontal 2G or 2F—The axis of the weld is approximately horizontal, but the type of weld determines the complete definition. For a fillet weld, welding is performed on the upper side of an approximately vertical surface. For a groove weld, the face of the weld lies in an approximately vertical plane (Figure 8-6).
- Vertical 3G or 3F—The axis of the weld is approximately vertical, Figure 8-7.
- Overhead 4G or 4F—Welding is performed from the underside of the joint, Figure 8-8.

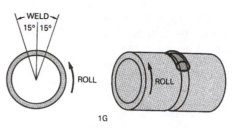

Figure 8-9 Pipe horizontal rolled position.

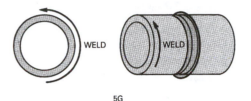

Figure 8-10 Pipe horizontal fixed position.

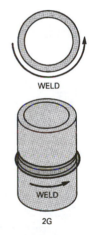

Figure 8-11 Pipe vertical position.

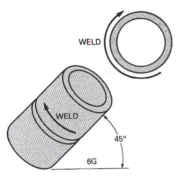

Figure 8-12 Pipe 45° inclined position.

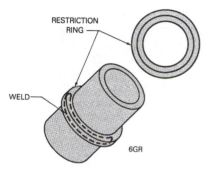

Figure 8-13 Pipe 45° inclined position with a restricting ring.

PIPE WELDING POSITIONS

The American Welding Society has divided pipe welding into five basic positions:

- Horizontal rolled 1G—The pipe is rolled either continuously or intermittently so that the weld is performed within 0°–15° of the top of the pipe, Figure 8-9.
- Horizontal fixed 5G—The pipe is parallel to the horizon, and the weld is made vertically around the pipe, Figure 8-10.
- Vertical 2G—The pipe is vertical to the horizon, and the weld is made horizontally around the pipe, Figure 8-11.

- Inclined with a restriction ring 6GR—The pipe is fixed at a 45° inclined angle, and a restricting ring is placed around the pipe below the weld groove, Figure 8-12 & 8-13.

METAL THICKNESS

The thickness of a metal is a major controlling factor in the selection of a joint design. On some thick sections it may be possible to make full penetration welds using a square butt joint (Figure 8-14). However, as the material becomes thicker, some method of edge preparation may be required if 100% joint penetration welds are required (Figure 8-15). The preparation of the edge for welding, grooving, or beveling may be on either one or both sides of the joint. The various edge shapes for this purpose include the following: beveled, V-grooved, J-grooved, and U-grooves (Figure 8-16). Some factors that

Figure 8-14 Full penetration weld on square butt joint.

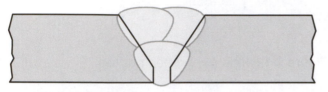

Figure 8-15 Multipass weld on thicker metal section that has been V-grooved for welding.

determine the shape of the edge preparation include type of metal, thickness, weld position, code or specification, and accessibility.

To help provide the required penetration, grooves or bevels may be cut into either one side or both sides of the joint (Figure 8-17). The edge may be prepared by grinding, flame cutting, gouging, sawing, or machining. Bevels and V-grooves are more easily cut on the parts before they are assembled. J-grooves and U-grooves may be cut either before or after assembly (Figure 8-18).

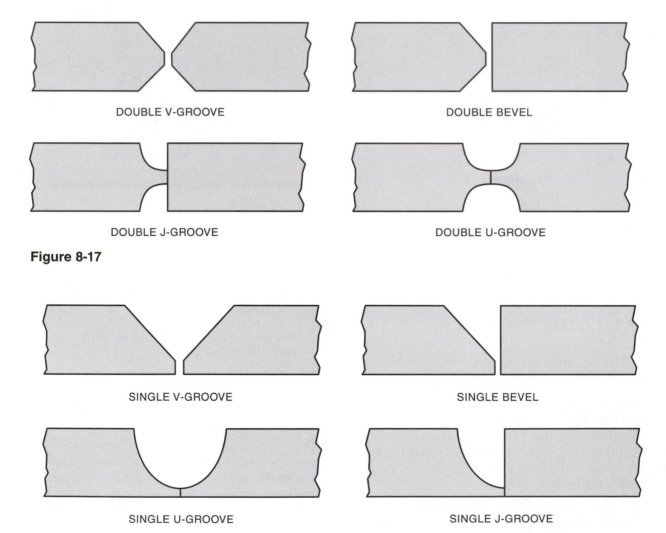

DOUBLE V-GROOVE

DOUBLE BEVEL

DOUBLE J-GROOVE

DOUBLE U-GROOVE

Figure 8-17

SINGLE V-GROOVE

SINGLE BEVEL

SINGLE U-GROOVE

SINGLE J-GROOVE

Figure 8-16

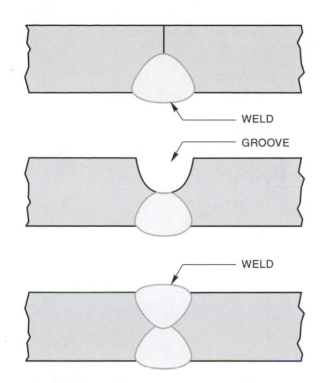

Figure 8-18 Back-gouging a weld to ensure 100% joint penetration.

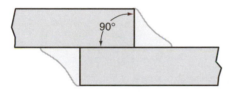

Figure 8-19

A 90° or squared edge is the only preparation used for lap joints. No strength can be gained by grooving the metal's edge for this joint (Figure 8-19).

If grooving is required, the decision to groove one or both sides of the plate is most often determined by joint design, position, code, and application. Plate in the flat position is usually grooved on only one side unless it can be repositioned so that the back side can be welded in the flat position also.

CODE OR STANDARDS REQUIREMENTS

The type, depth, angle, and location of the groove is usually determined by a code or standard that has been qualified for specific jobs. Organizations such as the American Welding Society, American Society of Mechanical Engineers, and the American Bureau of Ships are a few of the agencies that issue such codes and specifications. The most common code or standards are the AWS D1.1 and the ASME Boiler and Pressure Vessel (BPV) Section IX. The joint design for a specific set of specifications often must be prequalified, which means these joints have been tested and found to be reliable for the weldments for specific applications. The joint design can be modified, but the cost of having the new design accepted under the standard being used is often prohibitive.

WELDING SYMBOLS

Welding symbols enable a designer or an engineer to indicate clearly to the welder important details regarding the weld, including length, depth of penetration, height of reinforcement, groove type, groove dimensions, location, process, filler metal, strength, number of welds, weld shape, and surface finishing. Without welding symbols such detailed information would require lengthy written instructions to be included with the drawing.

The American Welding Society has established standards for welding symbols. Because of the complexity and diversity of these symbols, only the more common ones have been included in this chapter. Further information on all of the welding symbols is available in a booklet published by the American Welding Society and American National Standards. This booklet is entitled *Standard Symbols for Welding, Brazing, and Nondestructive Examination* (ANSI/AWS A2.4).

All of the components that make up a welding symbol are located on a horizontal reference line (Figure 8-20). An arrow extends from one end of the reference line and points to the location on the drawing where the weld is to be performed. A tail may be included on the opposite end of the reference line from the arrow. This tail is added to the symbol when it is necessary to include specific information (Figure 8-21), such as filler metal specifications, the number of weld passes, and pre- or postheat temperatures.

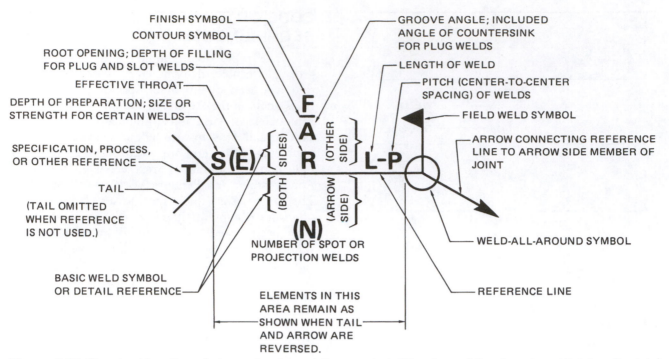

Figure 8-20 Standard location of elements of a welding symbol. (Courtesy of the American Welding Society)

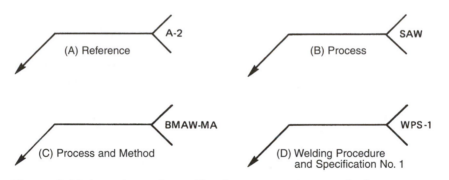

Figure 8-21 Locations of specifications, processes, and other references on weld symbols.

TYPES OF WELDS

Welds can be classified as follows: fillets, grooves, flange, plugs or slots, spot or projection, seam, back or backing, and surface. A specific symbol has been designed for each of these weld types and can be included on the reference line. The basic symbols are shown in Figure 8-22.

Although some of the symbols may closely resemble the weld shape that they indicate, they are only symbols and not pictorial representations of the weld to be performed. For example, the fillet weld symbol—a triangular shape—is always drawn with the vertical line on the right side and the sloped line on the left, regardless of the weld's actual configuration (Figure 8-23).

WELD LOCATION

The reference line is always drawn horizontally. Any symbol located on the top of this reference line is referred to as an *other-side* symbol. Any symbol located below the reference line is referred to as an *arrow-side* symbol. Both these terms refer to the side of the joint that the arrow touches (Figure 8-24).

The terms *arrow-side* and *other-side* do not change, regardless of the direction the arrow points. If the weld is to be deposited on the arrow side of the joint (near side), the proper weld symbol is placed below the reference line (Figure 8-25). If the weld is to be deposited on the other side of the joint (far side), the weld symbol is placed above the reference line. If a weld is to be deposited on both sides of the

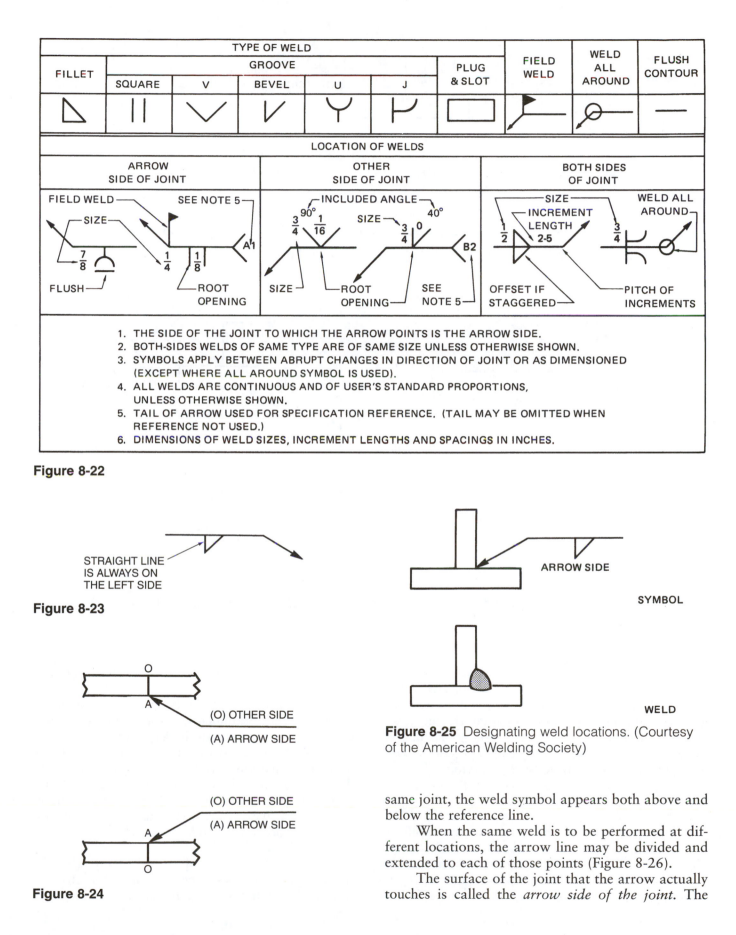

Figure 8-22

STRAIGHT LINE
IS ALWAYS ON
THE LEFT SIDE

Figure 8-23

(O) OTHER SIDE
(A) ARROW SIDE

Figure 8-24

ARROW SIDE

SYMBOL

WELD

Figure 8-25 Designating weld locations. (Courtesy of the American Welding Society)

same joint, the weld symbol appears both above and below the reference line.

When the same weld is to be performed at different locations, the arrow line may be divided and extended to each of those points (Figure 8-26).

The surface of the joint that the arrow actually touches is called the *arrow side of the joint*. The

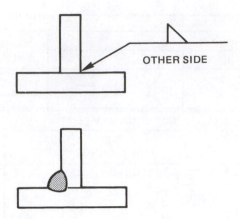

Figure 8-26 Designating weld locations. (Courtesy of the American Welding Society)

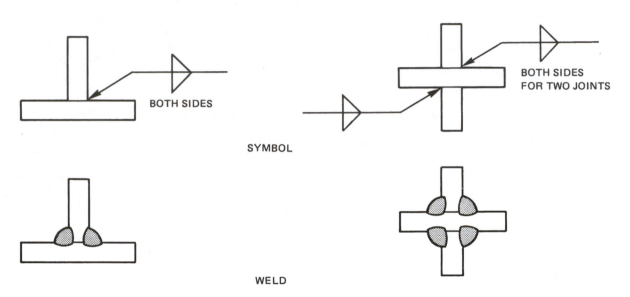

Figure 8-27 Designating weld locations. (Courtesy of the American Welding Society)

opposite side of the joint the arrow contacts is called the *other side of the joint*. On a drawing, when the joint is illustrated by a single line, this indicates that the weld joint extends outward behind the line (Figure 8-27). That surface represented by the line is called the arrow side of the joint, and the unseen back side is called the other side of the joint.

FILLET WELDS

Fillet weld dimensions are shown to the left of the welding symbol that indicates a fillet weld (Figure 8-28). If there is to be a weld on both sides of the joint, the fillet welding symbol appears both above and below the reference line; however, if both fillet welds are to be the same size, the dimension may be placed on only one side of the reference line. When both sides of the joint are to be welded with different size fillet welds, the dimensions for each weld appears on the appropriate side of the reference line.

The size of a fillet is given as the length of its leg. The dimension may be in fractions of an inch, decimals of an inch, or metric measurements and appear in parentheses to the left of the weld symbol. The leg of a fillet weld is the distance from the root measured horizontally and/or vertically to the toe of the weld (Figure 8-29).

The length of a fillet weld may also be included as part of its dimensioning. This information can be displayed in a number of ways on either

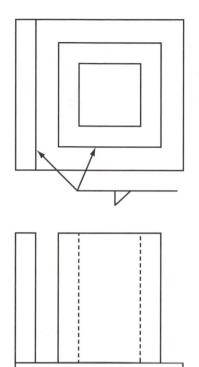

Figure 8-28

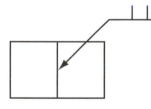

Figure 8-29

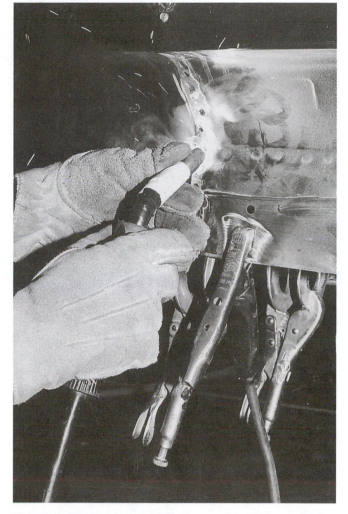

Figure 8-30 Making a MIG plug weld. (Courtesy of Larry Maupin)

Figure 8-31 Plug welding. (Courtesy of Larry Maupin)

the welding symbol or the drawing (Figure 8-30). When the length is indicated, it is shown to the right of the symbol (Figure 8-31). When the dimension is shown on the drawing, the arrow line intersects the appropriate dimension line that extends between two extension lines locating the weld's length (Figure 8-32).

For a variety of reasons, a fillet weld may not be made continuously along a joint. Intermittent welds are often used to control distortion and reduce the tendency for crack propagation. They also reduce weld cost by increasing production speed and

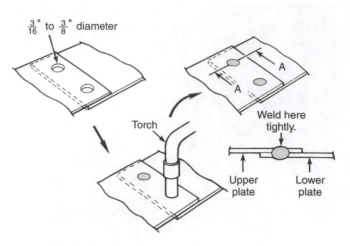

Figure 8-32 Steps in making a plug weld. (Courtesy of Nissan Motor Corp.)

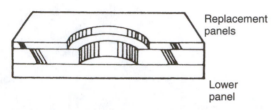

Figure 8-33 Welding two or more panels using the plug welding technique.

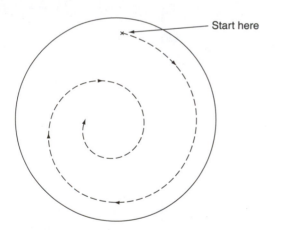

Figure 8-34 Start near the edge and spiral inward until plug is filled.

producing consumable supplies (Figure 8-33). The dimensions for an intermittent weld are given as the length and pitch. The length is the actual length of each individual weld. Thus a dimension of 3″ inches would indicate a series of 3″ long welds made in the joint (Figure 8-34). The pitch is the distance from the center of one weld to the center of the next. The pitch is often incorrectly thought to be the distance between each of the welds. To properly dimension an

Figure 8-35 Controls on typical MIG welder. (Courtesy of HTP America, Inc.)

Figure 8-36 Dimensioning the fillet weld symbol.

intermittent weld, both the length and pitch must be included. The first number represents the length, and the second number represents the pitch (Figure 8-35).

A combination of weld lengths can be incorporated on a single part. It may be necessary for an initial weld to be made from the corner or edge some distance along the joint and then intermittent welds to be made across the remainder of the joint. Figure 8-36 illustrates a joint using this method of weld configuration.

PLUG WELDS

The GMA plug weld can be used almost anywhere that a resistance spot weld was used. A plug weld has ample strength for welding load-bearing structural members and can be used on thin-gauge sheet metal.

Plug welding (Figure 8-37) is similar in function to spot welding; it is, however, welded through a hole. That is, a plug weld is formed by drilling or punching a hole in the outer panel being joined (Figure 8-38). The materials should be tightly clamped together. Holding the torch at right angles to the surface, put the electrode wire in the hole, trigger the arc briefly, and then release the trigger. Move the gun in a circle inside the hole so that the molten weld pool fills the hole and solidifies (Figure 8-39).

To make a plug, first punch or drill a hole through the top plate. When filling a hole, move the gun slowly in a circular motion around the edges of the hole (Figure 8-40), filling in the cavity. For smaller holes, it is best to aim the gun at the center and keep it stationary. A flat, gently raised bead gives a nice appearance and reduces the grinding or sanding operations.

Proper wire length is important in obtaining a good plug weld. If the wire protruding from the end of the gun is too long, the wire will not melt properly, causing inadequate weld penetration. The weld will

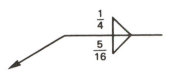

Figure 8-37 Dimensioning the fillet weld symbol.

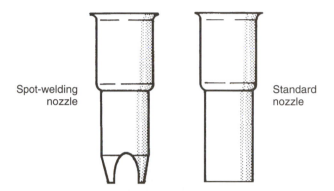

Figure 8-38 Note the notch in the spot-welding nozzle to let the heat and sparks out.

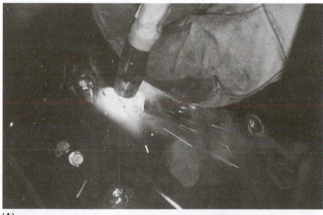
(A)

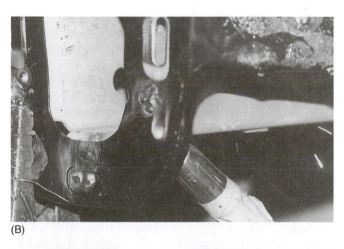
(B)

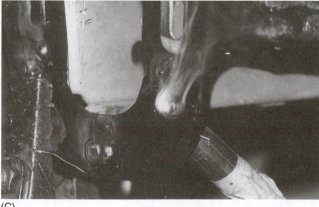
(C)

(D)

Figure 8-39 (A) Welding is started. (B) Back side heats up. (C) 100 percent penetration occurs. (D) Weld reinforcement can be seen as weld cools. (Courtesy of Larry Maupin)

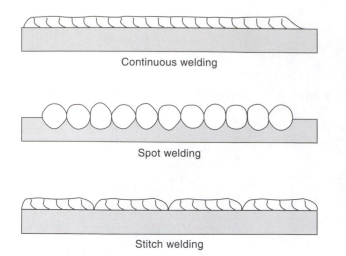

Continuous welding

Spot welding

Stitch welding

Figure 8-40 Finished weld looks continuous for both spot and stitch welds if done correctly.

Stitch welds

Figure 8-41 You can still see each weld bead in this lightly ground vertical stitch-welded rocker panel. (Courtesy of Larry Maupin)

improve if the gun is held closer to the base metal. Be sure the weld penetrates into the lower panel; there should be a round, or dome-shaped, protrusion on the underside of the metal as a good indicator of proper weld penetration.

Plug welds can also be used to join more than two plates. To do this, punch a hole in every plate except the lower one (Figure 8-35). The diameter of the plug weld hole in each additional panel should be smaller than the diameter of the plug weld hole on top. Likewise, if plates of different thicknesses are being joined, punch a larger hole in the thinner plate to ensure that the thicker plate is melted into the weld first. In addition, make sure the thinner plate is on top.

The welding symbol for a plug weld is a small rectangular box located on the reference line. Dimensioning information for the plug weld includes size, pitch, depth, and angle. The size refers to the diameter of the hole in the top plate. The pitch indicates the distance between the center of one plug weld and the center of the next. The depth is usually the same as the thickness of the top plate; however, the welds occasionally may not be filled flush with the top surface. That is the purpose of the depth dimension. The sides of a plug weld may have an angle other than 90°, which allows for better access to the base plate (Figure 8-36).

SPOT WELDS

GMA spot and plug welds are similar in appearance to the finished weld and application. Both of these techniques allow plates to be fused along an overlapping seam (Figure 8-41).

It is possible to spot weld with GMA welding equipment. Many GMA welding machines are now available with built-in timers that shut off the wire feed and welding arc after the time required to weld one spot (Figure 8-42). Some GMA welding equipment also has a burnback time setting that is adjusted to prevent the wire from sticking in the molten weld pool. The setting of these timers depends on the thickness of the workpiece. This information can be found in the machine's owner's manual.

Because spot welding usually requires more heat than making a continuous weld bead, make a test weld first to elect the right welder settings. Check the penetration of the spot weld by pulling the test weld apart. A good weld will tear a small hole out of the bottom piece, whereas a weak weld will break off at the surface. To get more penetration, increase the weld time or heat. To reduce penetration, reduce the time or heat.

For GMA spot welding, a special nozzle (Figure 8-43) must replace the standard nozzle. Once the gun is in place and the spot timing, welding heat, and backburn times are set for the given situation, the spot nozzle is held against the weld site, and the gun is triggered. The timed pulses of wire feed and welding current are activated very briefly, and the arc melts through the outer layer and penetrates the inner layer (Figure 8-44). Then the automatic shut-off goes into action, and, no matter how long the trigger is squeezed, nothing will happen. However, when the trigger is released and then squeezed again, the next spot pulse is obtained.

The quality of a GMA spot weld is difficult to ascertain because of varying conditions. Therefore, the plug weld or resistance spot welding tech-

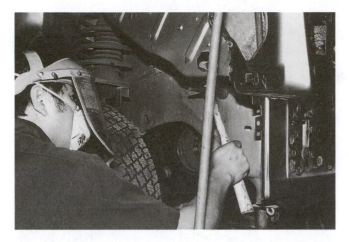

Figure 8-42 This auto body technician is wearing both a face shield and a respirator while working on zinc-coated metal. (Courtesy of Larry Maupin)

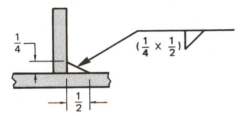

Figure 8-43 Dimensioning the fillet weld symbol.

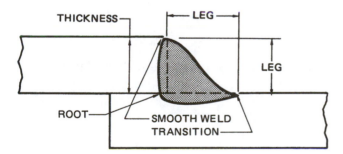

Figure 8-44 The legs of a fillet weld generally should be equal to the thickness of the bare metal.

nique described later is the preferred method on load-bearing members. However, the GMA lap spot technique is effective for quick welding of lap joints and flanges on thin-gauge nonstructural sheets and skins. Set the spot timer, and position the nozzle over the edge of the outer sheet at an angle slightly off 90°. This will allow contact with both pieces of metal at the same time. The arc melts into the edge and penetrates the lower sheet.

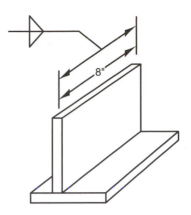

Figure 8-45

A GMA spot weld will have the spot welding symbol, a circle placed on the appropriate side of the reference line and the letters *GMAW* placed in the tail of the weld symbol (see Figure 8-45). The size of a spot weld may be dimensioned as either the diameter of the weld produced or its strength in either pounds (newtons) per square inch per weld or pounds per square inch per linear inch of weld (Figure 8-45). Additionally the pitch or center-to-center spacing of the weld may be given or the number of welds that are to be produced.

GROOVE WELDS

To ensure that the proper depth and penetration have been obtained, some joints are grooved before welding. There are several different types of grooves that can be made in one or both plates and on one or both of the joint's sides. Grooves are most often used to increase the joint's strength without reducing its flexibility.

A groove can be cut into the base metal in a variety of ways. A groove may be produced by using oxyfuel cutting torches, air carbon arc cutting, plasma arc cutting, machining, or sawing.

The various types of groove welds are classified as follows:

- Single-groove and symmetrical double-groove welds that extend completely through the members being joined. No size is included on the weld symbol (Figure 8-46).

- Groove welds that extend only part way through the parts being joined. The size as measured from the top of the surface to the bottom (not including reinforcement) is included to the left of the welding symbol (Figure 8-47).

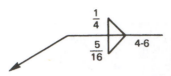

Figure 8-46 Dimensioning the fillet weld symbol.

Figure 8-49

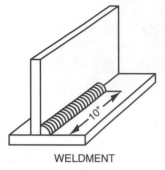

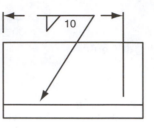

WELDMENT   DRAWING

Figure 8-47

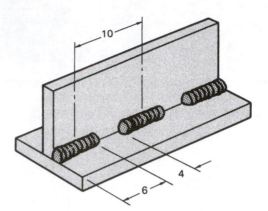

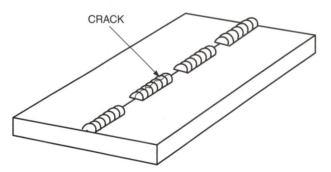

Figure 8-48 A crack that passes through one weld will not easily pass through the next weld.

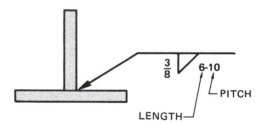

Figure 8-50 Dimensioning intermittent fillet welds.

- The size of groove welds with a specified effective throat is indicated by showing the depth of groove preparation with the effective throat appearing in parentheses and placed to the left of the weld symbol (Figure 8-48). The size of square groove welds is indicated by showing the root penetration. The depth of chamfering and the root penetration is read in that order—from left to right along the reference line.

- The root face's main purpose is to minimize the burn-through that can occur with a feather edge. The size of the root face is important to ensure good root fusion (Figure 8-49).

- The size of flare groove welds is considered to extend only to the tangent points of the members (Figure 8-50).

- The root opening of groove welds is the user's standard unless otherwise indicated. The root opening of groove welds, when not the user's standard, is shown inside the weld symbol (Figure 8-51).

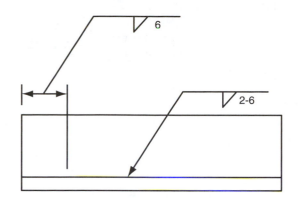

Figure 8-51

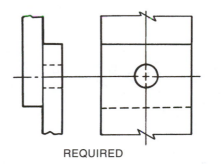

Figure 8-52 Plug weld.

BACKING

A backing (strip) is sometimes placed on the root side of a joint. The purpose of this backing strip for GMA welding is to both control the root face and prevent burn-through. The backing strip can be made out of the exact same material as the base metal, or it may be made from a material very similar to the filler metal. A backing strip may be used on butt joints, tee joints, outside corner joints, and piped joints (Figure 8-52).

Depending on the purpose of the finished product and any appropriate code or standard, the backing strip may be allowed to remain on the finished weld, or it may be removed. If the backing is to be removed, the letter *R* is placed in the backing symbol. The backing strip may be removed because it is often a source of stress concentration and a crevice in which trapped moisture can cause corrosion.

FLANGED WELDS

Flanged welds are sometimes used with thin-gauge sheet metal. Before welding begins, the edge of the sheet metal is bent to form a small flange (Figure 8-53). This flange controls joint distortion and burn-through and may serve as the filler metal for the completed weld. Flanged welds are most often performed in butt joints and outside corner joints.

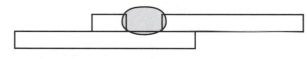

REQUIRED

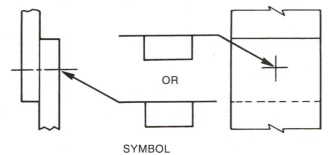

OR

SYMBOL

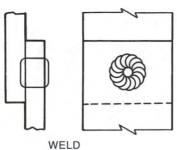

WELD

Figure 8-53

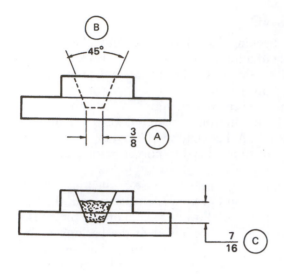

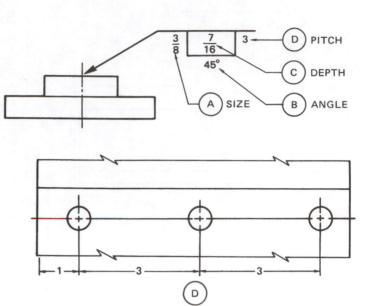

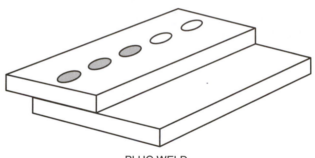

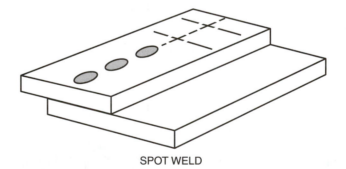

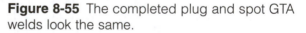

Figure 8-54 Applying dimensions to plug welds.

- Edge flange welds are shown by the edge flange weld symbol.
- Corner flange welds are indicated by the corner flange weld symbol.
- Dimensions of flange welds are shown on the same side of the reference line as the weld symbol and are placed to the left of the symbol (Figure 8-54). The radius and height above the point of tangency are indicated by showing both the radius and the height separated by a plus sign.
- The size of the flange weld is shown by a dimension placed outward from the flanged dimensions.

NONDESTRUCTIVE TESTING SYMBOLS

Because GMA welds are used in applications that require high-quality welds, they are often inspected before, during, and after the actual welding process. Therefore, a welder must become familiar with the nondestructive testing (NDT) standardized symbols, which use the same basic reference line and arrows as the welding symbol (Figure 8-55).

The symbol for the particular type of nondestructive testing to be performed is shown on the reference line. As with the welding symbol, the significance of having the NDT symbol above or below the reference line refers to arrow- or other-side of the joint. Symbols that are above the line indicate other

PLUG WELD

SPOT WELD

Figure 8-55 The completed plug and spot GTA welds look the same.

side; symbols that are below the reference line indicate arrow side; and those that are located on the reference line indicate that no preference exists for which side of the joint will be tested (Figure 8-56). Some tests may be performed on both sides of the

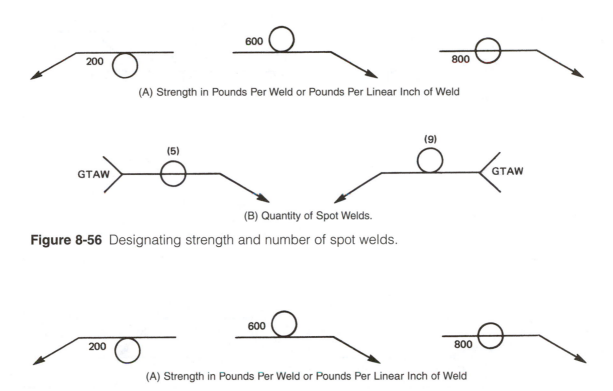

(A) Strength in Pounds Per Weld or Pounds Per Linear Inch of Weld

(B) Quantity of Spot Welds.

Figure 8-56 Designating strength and number of spot welds.

(A) Strength in Pounds Per Weld or Pounds Per Linear Inch of Weld

Figure 8-57 Arrow side 1/2-in. (13 mm) diameter GTAW spot weld.

weld surface; therefore, the NDT symbol will appear above and below the reference line.

In some cases two or more tests may be performed on the same joint. The first test usually is less expensive and is used as a screening test to determine whether more expensive testing should follow. By doing this the weld can frequently be repaired and brought up to standard before the complete bank is performed. Figure 8-57 shows several methods that are commonly used to combine NDT weld test symbols.

Dimensional information that may be included with NDT test symbols includes the length of weld to be tested and the number of locations the test is to be performed. The length dimension may be given to the right of the test symbol or can be shown by extension lines and an arrow line (Figure 8-58). The number of tests to be made is given in parentheses above or below the test symbol (Figure 8-59).

Both welding symbols and nondestructive testing symbols can be combined to form a single symbol (Figure 8-60). Combining the symbols can both inform welders of an impending postweld test and emphasize to them the importance of a particular weld's quality.

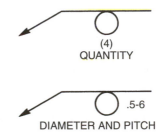

(4)
QUANTITY

.5-6
DIAMETER AND PITCH

Figure 8-58

Because an X ray is in essence the shadow of a weld captured on film, the angle used to expose the film is important. If the X ray is taken vertically through a joint, there is no depth perception available to the technician reading the X ray to determine exactly where discontinuities may appear within the weld. If the angle is too sharp, the discontinuity showing on the film may also be skewed excessively, making size determination difficult. The engineer or designer of the weldment can include the desired exposure angle as part of the radiographic testing symbol.

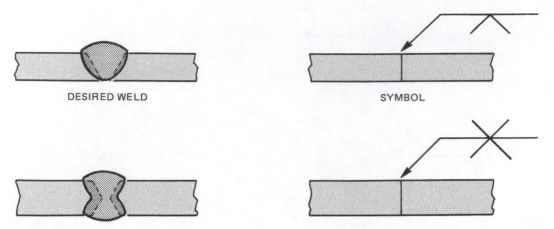

Figure 8-59 Designating single- and double-groove welds with complete penetration. (Courtesy of the American Welding Society)

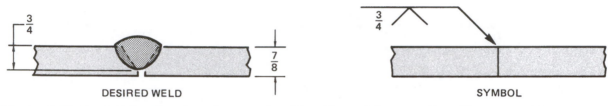

Figure 8-60 Designating the size of grooved welds with partial penetration. (Courtesy of the American Welding Society)

REVIEW QUESTIONS

1. What is a welding symbol?
2. What weld position allows for the largest possible weld pool to be controlled?
3. What is a major controlling factor in the selection of a joint design?
4. What information is included in the welding symbol?
5. What can welds be classified as?
6. What is an arrow-side symbol?
7. When both sides of the joint are to be welded with different size fillet welds, the dimensions for each weld appears where?
8. Where can a GMA plug weld be used?
9. What is an important factor in obtaining a good plug weld?
10. What is the welding symbol for a plug weld?
11. When is a groove used?
12. What is the purpose of the backing strip for GMA welding?
13. Where are flanged welds most often performed?
14. What happens if an X ray is taken vertically through a joint?

CHAPTER 9

BLUEPRINT READING

Mechanical drawings have been called the universal language because they are produced in a similar format worldwide. There are only a few differences in the arrangement of the views in a drawing (Figure 9-1). These slight differences in arrangement are minor, and once you understand how to place the various views in the drawing, you will have no difficulty interpreting them correctly. Such interpretation is made easier because the method for representing a view on both types of drawings is the same as the method for dimensioning.

A set of drawings contains all of the information required to produce a particular part, fabrication, or finished product. This set of drawings may consist of a single sheet with one or more drawings, or it may consist of multiple sheets as may be required for larger, more complex parts. Pages that may be included on a larger set of drawings are the title page, pictorial, assembly drawing, detailed drawing, and various views (Figure 9-2).

In addition to the specifications and dimensions included on the drawing, you may also find a bill of

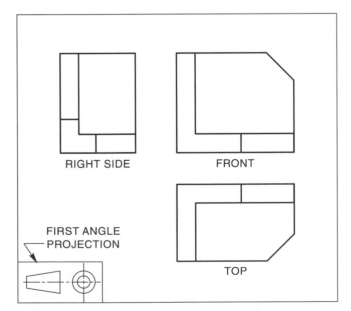

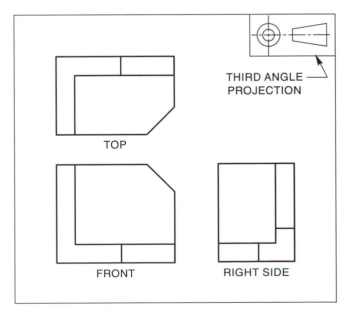

Figure 9-1 Two methods for rotating drawing views.

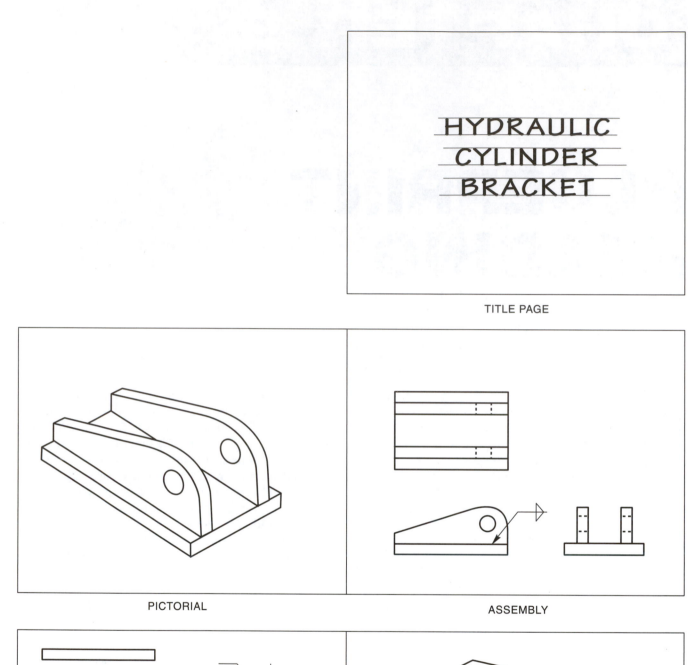

TITLE PAGE

PICTORIAL

ASSEMBLY

DETAIL

EXPLODED

Figure 9-2 Drawings that can make up a set of drawings.

BILL OF MATERIALS				
Part	Number Required	Type of Material	Size Standard Units	Size SI Units
Base	1	Hot Roll Steel	1/2" × 5" × 8"	12.7 mm × 127 mm × 203.2 mm
Cleat	2	Hot Roll Steel	1/2" × 4" × 8"	12.7 mm × 101.6 mm × 203.2 mm

Figure 9-3

materials and a title block. The bill of materials is a detailed list of the materials, including grades of metal, thicknesses, and sizes for each component or item needed to build the weldment (Figure 9-3).

The title box usually appears in one corner of the drawing or on one edge and contains general information about the part and specific information about the drawing. General information about the part includes its name, the name of the individual or business that ordered the part, and tolerances. Information about the drawing may include the scale and the date of the drawing, the name of the person who made the drawing, the page number, the number of pages in the set, and, if this is a revised drawing, the effective date of the revision.

LINES

The drawing consists of a variety of types of lines, each of which has a specific meaning. The lines are collectively known as the *alphabet of lines* (Figure 9-4). A description of the most commonly used lines and their purpose is as follows:

- Object Line—An object line is a solid dark bold line that shows the shape of an object, including the outline and other significant features such as holes, slots, the edges of different pieces that make up the part, and, on flat parts, any corners created by a change in angle. On round or circular parts, the object line illustrates the maximum diameter or radius. This is true even though the part may look flat as seen in one view (Figure 9-5). In such a case, you must look at another view to determine the exact shape of the object.

- Hidden Line—A hidden line is a medium-weight series of short dashes that can range from approximately 1 1/4" to 3/8" in length.

The space between the dash may be from 1/16" to 1/8". The dashes and spaces used on the drawing should be consistent. Hidden lines represent the same thing that object lines represent except that the surface of extent of the curb surface occurs below or behind a solid surface. Hidden lines therefore represent notches, grooves, holes, and so on that are to be made in the back or bottom of the part that would not be visible when looking at the object from this direction.

- Center Line—A center line is a thin, fine broken line made up of longer line sections with shorter dashed line sections (Figure 9-6). The long line section may be from 1/2" to several inches in length. The short dashed sections may range from 1/4" to 1/2" in length. Spacing between each of these lined sections may range from 1/8" to 1/4". The center line is used to show the centers of holes, curves, or symmetrical parts. Center lines may also be extended to show dimensioning (Figure 9-7).

- Extension Line—An extension line is a thin, fine line that is drawn as an extension of an object line. There must be a 1/16" to 1/8" gap between the end of the extension line and the object line it extends from (Figure 9-8). Extension lines are used for dimensioning and should extend 1/8" to 1/25" beyond the last dimension line. Extension lines may be used to represent the extent of a curved surface or circular object (Figure 9-9).

- Dimension Lines—Dimension lines are medium-weighted lines that locate the exact spots between which the dimension refers (Figure 9-10). Dimension lines may extend from an extension line or object line. The exact point they reference is identified by

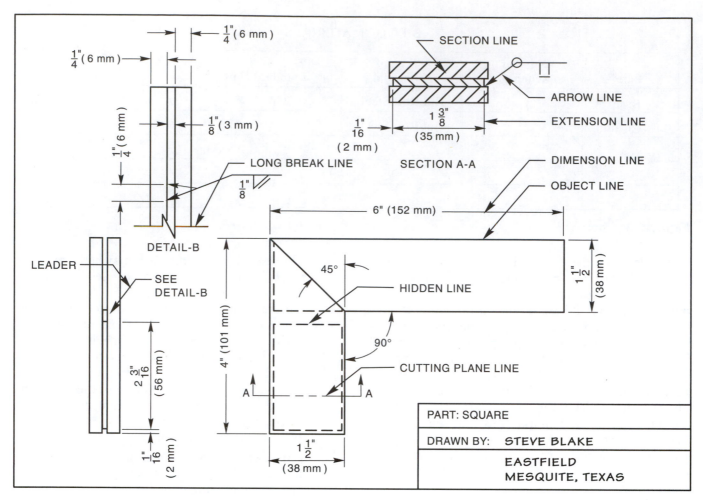

Figure 9-4

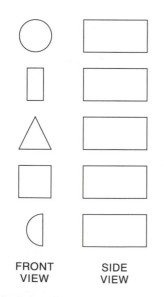

FRONT
VIEW

SIDE
VIEW

Figure 9-5 All of the side views would look the same
for each of these different front views.

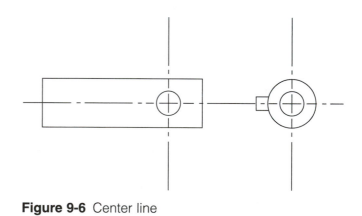

Figure 9-6 Center line

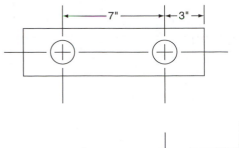

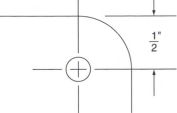

Figure 9-7 Using center lines for dimensioning.

Figure 9-8

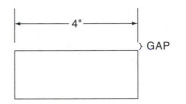

Figure 9-9 Extension lines.

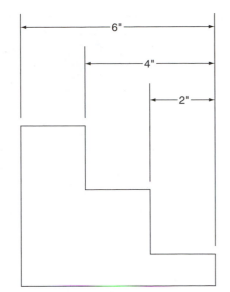

Figure 9-10 Dimension lines.

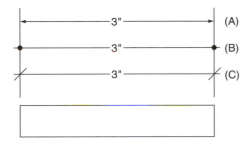

Figure 9-11 The end of a dimension line can have (A) an arrow, (B) a dot, or (C) a slash.

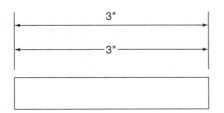

Figure 9-12

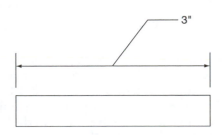

Figure 9-13

placing an arrow, a dot, or a slash on the dimension line (Figure 9-11). The dimension line may have a short break in the middle where the dimension is placed, or the dimension may be written just above or below the dimension line (Figure 9-12).

- Cutting Plane Line—A cutting plane line is a bold line made up of a series of long and short dashed lines. The long dashed line can range from 1/2″ to 1″ in length, and the short dashed line can range from 1/4″ to 3/8″ in length. Spaces between the lines ranges from 1/8″ to 1/4″. Each end of the line has a short line drawn at a right angle to the main line with arrows drawn at their end (Figure 9-13). There is usually a letter drawn at each end of the

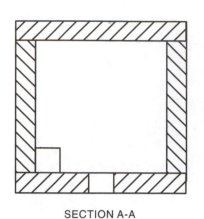

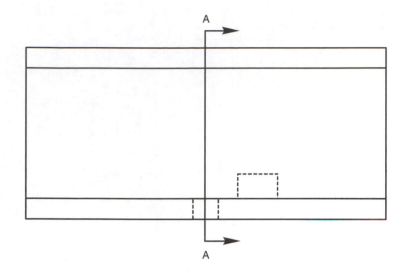

SECTION A-A

Figure 9-14

cutting plane line. This letter designates the detailed drawing that its particular cutting plane line locates. Cutting plane lines conceptually "open up" the inside of a part to show details that could not otherwise be clearly illustrated (Figure 9-14).

- View Line—A view line is drawn exactly the same as the cutting plane line except that shows surface details and not interior or hidden details. Letters attached to the viewline refer to a detailed drawing that can be seen clearly only from looking at the object in this specific direction.

- Section Lines—Section lines are thin, fine lines drawn parallel to each other but at an odd angle to the object lines. This angle will change from part to part within the sectional drawing (Figure 9-15). A large number of different materials can be represented by the various spacing of section lines (Figure 9-16). However, most draftspersons use the symbols for iron or steel when making sectional drawings, regardless of the actual material the parts are made from. Section lines illustrate only the imaginarily cut surfaces that are identified by a cutting plane line.

- Leader or Arrow Lines—These are medium-weighted lines that have an arrowhead drawn on one end. Leader and arrow lines point to a specific location that a note, specification, or dimension refers to. Arrow lines, in conjunction with welding symbols, locate the joint or surface to be welded.

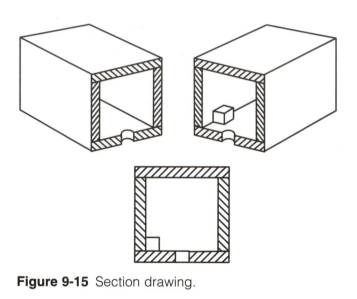

Figure 9-15 Section drawing.

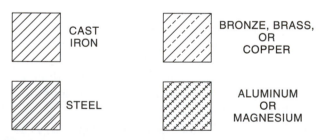

Figure 9-16 Types of section lines.

- Break Lines—Two types of break lines may be used on a drawing. One is a bold, straight line that has intermittent zig-zags drawn along its length (Figure 9-17). The other is a bold, free-hand irregular line (Figure 9-18). Both of these

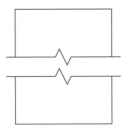

Figure 9-17 Break line.

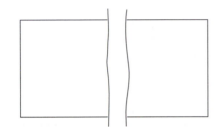

Figure 9-18 Freehand break line.

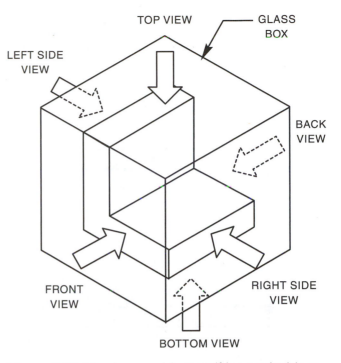

Figure 9-19 Viewing an object as if it were inside a glass box.

lines illustrate a portion or segment of a part that is not included on the drawing. Break lines are sometimes used to remove central portions of an object when no details are required for the fabrication of the part. Such removals allow the draftsperson to fit the part on the drawing page while using a larger scale for the drawing.

TYPES OF DRAWINGS

Two general categories of drawings are used in the welding industry. The majority of all drawings are *orthographic* projections, sometimes referred to as mechanical drawings. These drawings are made as if you were looking through the sides of a glass box at the object (Figure 9-19). The object then appears as if you had traced its shape on the side of the glass box and then unfolded the box, laying it flat. If you were to trace the fixed sides of an object and lay them out as if the box had been unfolded, they would appear as shown in Figure 9-20.

The other type of drawing used to represent an object is called a *pictorial*. These drawings are more realistic and depict the object as it would appear if you were looking at it. Pictorial drawings can be divided into three major types: *isometric, cavalier,* and *angular* perspectives, (Figure 9-21). Angular perspective drawings are the most realistic and are sometimes used to illustrate to a prospective customer what a finished product will look like. They

are seldom used by the welder for actual fabrication purposes.

Projection Drawings

Although up to six views can be included in a projection drawing, seldom are more than three different views used. For example, when looking at a car, the right and left sides for all practical purpose would appear on a drawing as mirror images of each other. Therefore, it would not be necessary for the draftsperson to have included both the right and left sides since one view would not contain significant details not easily obtained from the other. The three most common views used in orthographic projections are the front view, right-side view, and top view (Figure 9-22). In some cases, when relatively simple objects are being illustrated, it may be necessary to use only one or two of these views.

The front view is not necessarily of the front of the object. The side selected to represent the front view should be the side that best depicts the overall shape of the object. As an example, the "front" view of a car or truck would probably be of the side of the vehicle, which would show more about the vehicle than the front. The front of many cars, light trucks, station wagons, and vans all appear similar; therefore, the side view of the vehicle would be necessary to determine the actual type of vehicle.

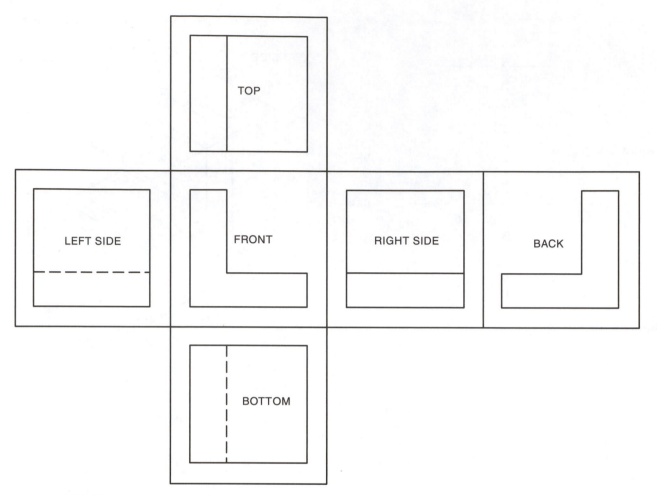

Figure 9-20 The arrangement of views of an object if the glass box were unfolded.

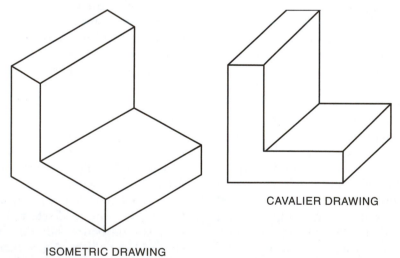

ISOMETRIC DRAWING

CAVALIER DRAWING

Figure 9-21 Pictorial drawing types.

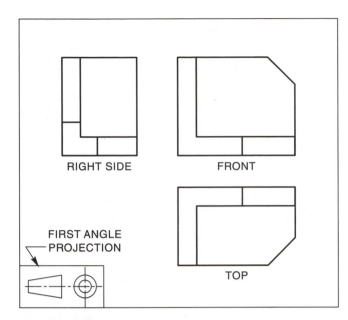

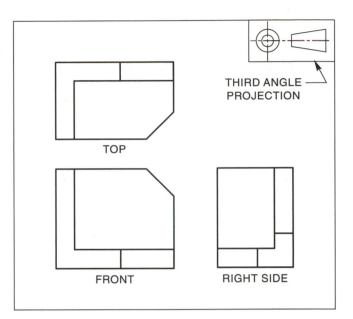

Figure 9-22 Two methods for rotating drawing views.

Special Views

Special views may be included on a drawing to help describe the object so it can be made accurately. Special views on some drawings may include the following:

- Section View—A section view is what you would see if you were to saw away part of the object to reveal internal details (Figure 9-23). This is done when the internal details would not be as clear if they were shown as hidden lines. Sections can be either fully across the object or just partially across it. The imaginary cut surface is set off from other noncut surfaces by section lines drawn at an angle on the cut surfaces. On some drawings the type of section line used depicts the type of material the part was made with. The location of this imaginary cut is shown by a cutting plane line (Figure 9-24).

- Cut-Aways—A cut-away view shows details within a part that would be obscured by the part's surface. Often a freehand break line is used to outline the area that has been "removed" to reveal the inner workings.

- Detail Views—A detailed view is usually an external view of a specific area on the part. Detail views show small details of an area on a part without having to draw the entire part larger. Sometimes only a small portion of a view is important, and this area can be shown in a detailed view, which can be drawn at the same scale or larger if needed. By showing only what

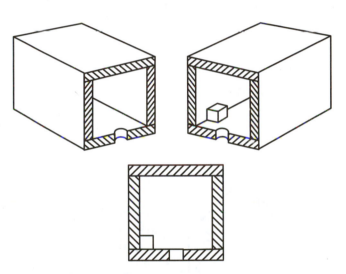

Figure 9-23 Section drawing.

is needed within the detail, the part drawn can be clearer and not require such a large page.

- Rotated Views—A rotated view can show a surface of the part that would not normally be drawn square to any one of the six normal view planes. If a surface is not square to the viewing angle, then lines may be distorted. For example, a circle when viewed at an angle looks like an ellipse.

Dimensioning

Drawings are two-dimensional representations of three-dimensional objects. In other words, on any

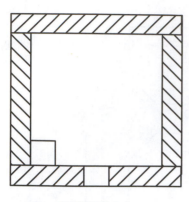

SECTION A-A

Figure 9-24

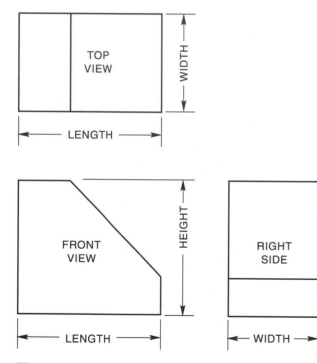

LENGTH

Figure 9-25

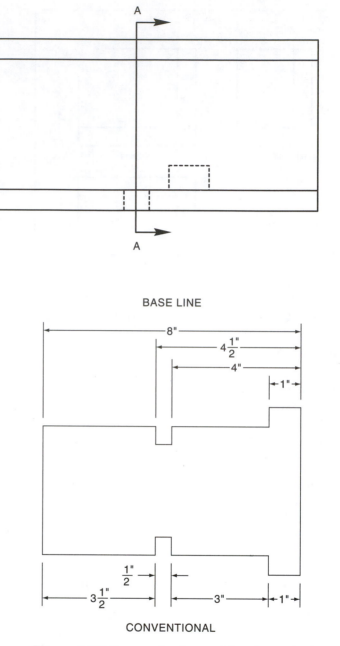

BASE LINE

CONVENTIONAL

Figure 9-26 Two methods used for dimensioning.

single view of an orthographic projection, only two dimensions may be illustrated. The length dimension, for example, can be found on the front and top views. The height dimension can be found on the front and right-side views. The width dimension can be found on the top and right-side views (Figure 9–25). It is necessary therefore to look at other views in order to locate all of the dimensions that would be required to build an object. Dimensions on a drawing may be given conventionally or from a base line (Figure 9-26). Conventional dimensioning refers to a method in which each dimension is given from a point, and that

point's dimension is used as a reference for other points. With this type of dimensioning, it may be necessary to add together dimensions in order to locate some items on a part. For example, in Figure 9-27, in order to locate the center of the hole, the welder must add together the dimension from the end of the bracket to the brace and then the distance from the brace to the hole.

Baseline dimensioning uses an edge, side, or major component as a baseline, and all dimensions' distances extend from that single point. With baseline dimensioning, the welder does not have to add up

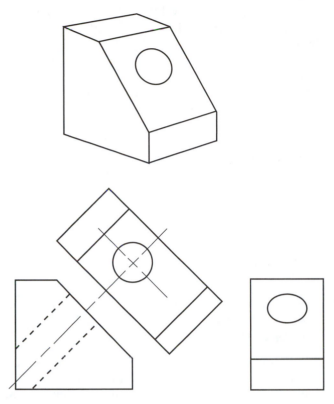

Figure 9-27 Notice that the round hole looks misshapen, elliptical, in the right side view.

dimensions to locate points. (A drawing may become overly cluttered and thus confusing if a large number of dimensions is given.

In both dimensioning systems, it is often necessary to locate the dimension distance for any given item on a single view by referring to the item in another view. In Figure 9-27, in order to locate the center of the hole on the front view, you would have to locate the length dimension in the top view and the height dimension in the side view. Such cross-referencing to locate all of the dimensions is common. If all of the dimensions of every component were shown on each and every view, the drawings would become quite cluttered. Therefore, only the dimensions required to accurately produce the part are given—and usually only once.

If, after a diligent search, you cannot locate a dimension on a drawing, do not try to obtain the dimension by measuring the drawing itself. Even if the original drawing was made very accurately, the paper it is drawn on will change sizes with changes in humidity. Copies of the original drawing will never be the exact same size. The most acceptable way of determining a missing dimension is to contact the person responsible for the drawing.

THE METRIC SYSTEM

During the 1960s the United States committed itself to changing from the standard U.S. measurement system to the metric system. As global markets and international competition developed in manufacturing, in the aerospace industry, in transportation, and in the computer, aircraft, and robotics industries, the metric system evolved into the International System of Units (SI). This measurement system is used worldwide to improve the function of a global economy. The welding industry in the United States has been slowly moving toward using this official, modernized system.

GMAW Welders and SI

Today's welders frequently use the SI units of measurement. You will need to know and use this metric system in the workplace. Fortunately, it is not difficult to learn and use.

International System of Units (SI)

SI uses seven basic units of measurement:
1. Electrical current
2. Length
3. Luminous intensity
4. Mass
5. Substance
6. Temperature
7. Time

SI also uses a few supplementary units for measuring angles in two and three dimensions.

SI Basic Units

The basic SI units you will apply in everyday use are these:

Meter–Basic unit of length

Gram–Basic unit of mass (weight)

Liter (litre)–Basic unit of capacity

Keep in mind that these are not all the standards of the SI metric system, but they are basic units to apply to common uses.

The three main SI standards are the meter, the kilogram, and the liter. All basic units, including these, are based on the SI standards.

SI Prefixes

There are six main prefixes used with basic SI units of measurements:
1. Kilo–1,000 (one thousand)
2. Hecto–100 (one hundred)
3. Deca–10 (ten)

4. Deci–1/10 (one-tenth)

5. Centi–1/100 (one-hundredth)

6. Milli–1/1000 (one-thousandth)

With the prefixes and basic units on a scale from larger to smaller, a factor of ten (10) exists between each step.

Keep in mind that you never use a prefix without using a basic unit with it. A basic unit, however, may be used alone.

Using a prefix is simple. If you want to name a multiple of a basic unit, use a prefix.

Example: kilometer

1. Kilo is the prefix (meaning 1,000).

2. Meter is the basic unit of length.

3. Kilometer represents a length of 1,000 meters.

If you want to name a decimal part of a basic unit, use a prefix for one of the decimal parts and the basic unit.

Example: milligram

1. Milli is the prefix (meaning 1/1000).

2. Gram is the basic unit of mass.

3. Millimeter represents a mass of one-thousandth of a gram.

SI Symbols

The symbols for the SI basic units and prefixes are these:

m = meter

k = kilo

d = deci

g = gram

h = hecto

c = centi

l = litre

da = deca

m = milli

Using the Welder's SI Stairway

When the SI basic units are known and you understand the prefixes and symbols, converting from smaller to larger units is easy.

Set the prefixes in order on the welder's SI "stairway" steps. Refer to stepping "up" or "down" on the stairway. Each step represents a factor of 10. As you go up one step, you move to the SI unit that is ten times greater than the one below it.

Example: As you go up one step, 1 meter becomes 10 meters. Go up another step, and the original 10 meters becomes 100 meters. Move the decimal point to the left when going up the welder's SI stairway.

As you travel down the welder's SI stairway, the SI unit is one-tenth as large as (or 10 times smaller than) the one above it.

Example: As you go down, 1 meter becomes smaller. Moving down one step would be 1 decimeter, or .1 (1/10) meter. Moving down another step would be 1 centimeter, or .01 (1/100) meter.

SI ALERT

If the conversion is from small units to larger units, STEP UP the stairway. If the conversion is from larger to smaller units, STEP DOWN the stairway.

When measuring in SI units, express your measurement in only one unit.

Example: 9.63 m–This measurement should be expressed as 9.63 m *not* 9 m 63 m, or in any other form.

SUMMARY

Drawings are often used by more than one person during fabrication. So to avoid possible confusion by anyone, do not write or do calculations on the drawings. The better care you take of these drawings, the easier it will be for someone else to use them. Always keep drawings clean and well away from any welding.

REVIEW QUESTIONS

1. What information does a set of drawings include?

2. What information does a title box contain?

3. What does the center line show?

4. What is the cutting plane line used for?

5. What are the two types of drawings?

6. How many views can be included in a projection drawing?

7. Where are height dimensions found?

8. What are the basic SI units you will apply in everyday use?

9. What are the three main SI standards?

10. What does *kilometer* represent?

CHAPTER 10

LAYOUT AND FABRICATION

Depending on the size of the shop and the type of work being performed, most welders will be required to lay out and fabricate components they are going to weld. In large shops the laying out and fabricating of virtually all of the components is performed by persons other than the welder. This frees up the time of the skilled welders and enables them to produce more welds. Also in large shops the parts may be preassembled and tacked ready for welding. Even under these working conditions, from time to time these welders will be asked to lay out and fabricate specialty parts.

Layout is the process of transferring the drawing specifications of parts from paper to the metal that they will be fabricated from. Once laid out, the parts may be cut out using any number of processes such as plasma arc cutting, oxyfuel cutting, shearing, machining, or laser cutting.

Fabrication refers to the assembling of parts into the correct positions as specified by the drawings. Some assemblies may be done by hand; others may use jig fixtures, clamps, or other means to help accurately position each component that will make up the weldment.

To develop your welding fabrication and layout skills, you can treat each welding practice plate as if it were a part being assembled for production welding.

LAYOUT

In the layout process, lines from drawings, sketches, or concepts within the welder's mind are transferred onto the material that will be cut, bent, grilled, punched, assembled, and welded. During layout, the welder will transfer dimensions and specifications located on the drawing to the material to be used for fabrication. Not all layout lines are used to locate cuts to be made on the material. Some layout lines show the positions for locating parts during the assembly. Yet other lines are used only to help in the layout process; these line are referred to as *construction lines* (Figure 10-1). The purpose of each line drawn on the material must be clearly marked to prevent mistakes during fabrication.

The accuracy and squareness of all lines is essential. Fabricated parts can be no more accurate than the lines used during assembly and fabrication.

Lines may be marked with a soapstone or chalk line, scratched with a metal scribe, or punched with a center punch (Figure 10-2). Other items that may be used for drawing on the material (such as pencils, pens, felt-tip markers, and paint pens) may contribute to contamination of gas tungsten arc welds.

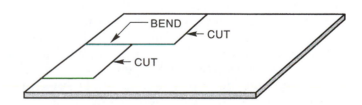

Figure 10-1 Identifying layout lines to avoid mistakes during cutting.

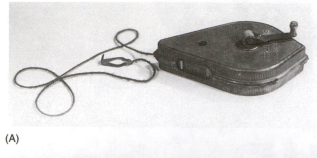

(A)

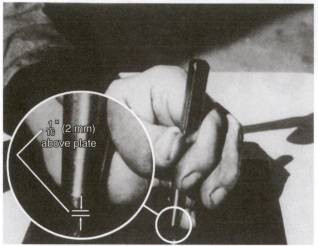

$\frac{1}{16}$" (2 mm) above plate

(B)

Figure 10-2 (A) Chalk line reel. (B) Holding the punch slightly above the surface allows the punch to be struck rapidly and moved along a line to mark it for cutting.

These items are frequently used during the layout and fabrication of materials that are to be welded using other processes, such as shielded metal arc welding, flux core welding, and others in which fluxes that are part of the process can remove small contaminations that might result from these marks.

Soapstone is available in either flat or round-cut plates. To draw accurate lines, you must sharpen the soapstone properly (Figure 10-3).

A chalk-line reel contains powdered chalk and a cotton frame. Chalk lines can make long, straight lines on metal and work best on large jobs. Powdered chalk is available in a variety of colors including red, green, blue, and white. As the string is pulled from the chalk line, powdered chalk adheres to the string. The string straightens as it is pulled tight between two marks. Lifting the center of the string and allowing it to snap back against the metal results in a line of powdered chalk sticking to the metal surface (Figure 10-4).

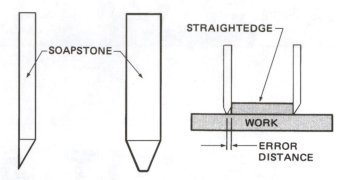

Figure 10-3 Proper method of sharpening a soapstone.

Part Layout

Always start a layout as close as possible to a corner or edge of the material. This will both aid in layout accuracy and help reduce wasted material. Often you can take advantage of a preexisting straight edge on the stock and use it as one surface of the part you are fabricating. An existing edge also aids in aligning tools such as squares and bevels.

It is often easy to mistakenly cut, bend, or locate a part on the wrong line. In welding shops one person may lay out the parts, and others may cut them out or assemble them. However, even if the same person does all of these jobs, it is still possible to use the wrong line because of such situations as a restricted view through cutting goggles or confusion caused by a large number of lines during layout.

To avoid such mistakes, you should identify lines as to whether they are for cutting, bending, drilling, or assembly locations. Common methods used to identify lines (such as those that are not to be cut) is to mark them with an *X*. Lines may also be identified by writing directly on the part. Additionally, to ensure the accurate cutting out of parts, sometimes the scrap side (where the kerf is to be removed during cutting) should be identified (Figure 10-5). Any lines that were used for constructing the actual layout lines should be erased completely or clearly marked to avoid confusion.

Some shops have their own shorthand methods for identifying layout lines, or you may develop your own system. If a mistake occurs during cutting out or fabrication, check with your welding shop supervisor to see what corrective steps you should take. One advantage of most welding assemblies is that such errors may be repairable by welding. There are often prequalified procedures established for just such events, so check before you decide to scrap a part.

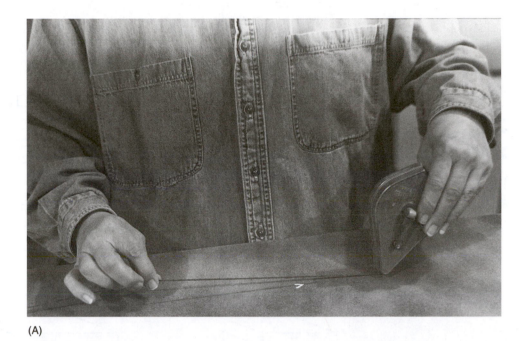

(A)

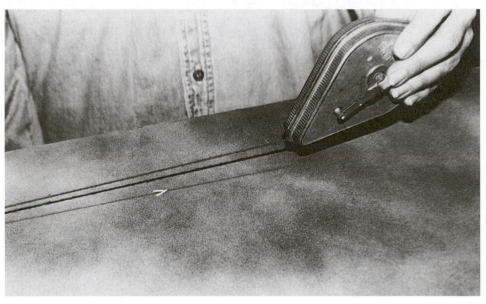

(B)

Figure 10-4 (A) Pull the chalk line tight and then snap it. (B) Check to see that the line is dark enough.

Figure 10-5

The process of laying out a part may be affected by the following factors:

- *Material shape*–Figure 10-6 lists the most common shapes of metal used for fabrication. Flat stock such as sheets and plates are easiest to lay out, and the most difficult shapes to work with are pipes and round tubing.

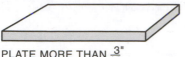
PLATE MORE THAN $\frac{3"}{16}$

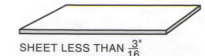
SHEET LESS THAN $\frac{3"}{16}$

EXPANDED SHEET

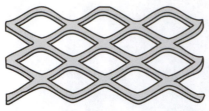
STANDARD

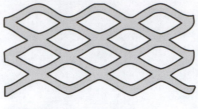
FLATTENED

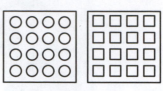
PERFORATED SHEET

STRUCTURAL SHAPES

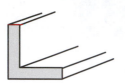
ANGLES

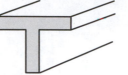

TEES

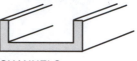
CHANNELS

STANDARD BEAMS

H BEAMS

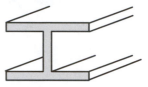
WIDE FLANGE

PIPE

STANDARD SCHEDULE 40

EXTRA STRONG
SCHEDULE 80

DOUBLE EXTRA STRONG
SCHEDULE 180

TUBING

ROUND

SQUARE

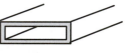
RECTANGULAR

BAR STOCK

ROUND

SQUARE

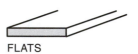
FLATS

HALF-ROUNDS

HALF-OVAL

ZEE BAR

HEXAGON

OCTAGON

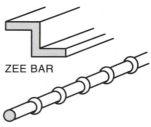
REINFORCING BAR

Figure 10-6 Standard metal shapes. Most are available with different surface finishes such as hot-rolled, cold-rolled, or galvanized.

- *Part shape*–Parts with square and straight cuts are easier to lay out than are parts with angles, circles, curves, and irregular shapes.
- *Tolerance*–The smaller or tighter the tolerance that must be maintained, the more difficult the layout.
- *Nesting*–The placement of parts together in a manner that will minimize waste is called *nesting*.

Parts that are square or have straight edges are the easiest to lay out. Square lines can be laid out using tools such as framing squares, combination squares, tri-squares, or other such devices. They may also be made by simply measuring equal distances from existing right-angle corners on the stock.

Straight or parallel cuts can be made by using a combination square and a piece of soapstone. Set the combination square to the correct dimension, and drag it along the edge of a plate while holding the soapstone at the end of the combination square blade (Figure 10-7).

Circles, Arcs, and Curved Lines

Circles, arcs, and curves can be laid out by using either a compass or circle template (Figure 10-8). The *diameter* is usually the dimension given for holes or other such circular parts. The *radius* is usually the dimension given for arcs and curves. The center point of the circle, arc, or curve may be located using dimension lines and center lines (Figure 10-9). Curves and arcs that are to be made tangent to another line may be dimensioned with only their radius (Figure 10-10).

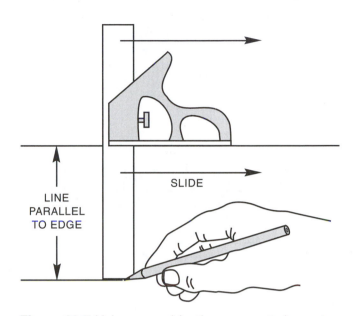

Figure 10-7 Using a combination square to lay out a strip of metal.

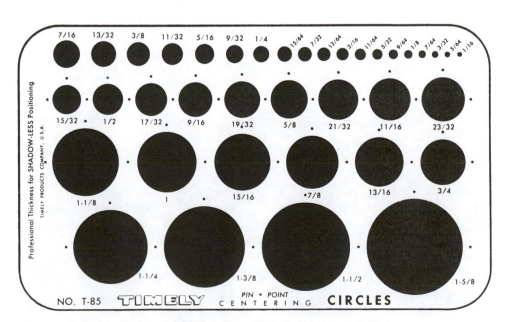

Figure 10-8 Circle template. (Courtesy Timely Products Co.)

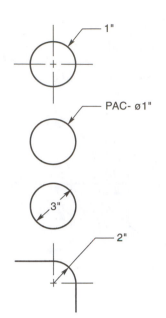

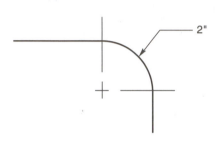

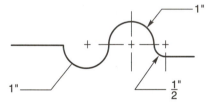

Figure 10-9

Figure 10-10

Angled Lines

Straight line that not drawn at a 90° angle or square can be dimensioned using either their angle or the points between which the lines will be drawn (Figure 10-11). Triangles that have 30°-, 60°-, 90°-, or 45°-, 90°-angles are available for such layouts. Odd angles can be located by using a protractor (Figure 10-12).

On large parts, angles are more accurately laid out if dimensions are given for their points. Even the slightest angular error as marked using a protractor can result in an inaccurate part if the line is to be extended several feet (Figure 10-13). The interior angles of all triangular parts must add up to 180° (Figure 10-14).

Nesting

Laying parts so that they are clustered tightly together, thus eliminating as much scrap as possible, is called *nesting* (Figure 10-15). Odd-shaped and unusual sizes of parts produce the largest amount of scrap and present the greatest difficulty for nesting. It is often necessary to try several layouts before determining the pattern that will minimize scrap. Sometimes these trial-and-error efforts can be accomplished using a pencil and paper.

Computer programs can aid in layout. Some automated cutting machines can use these computer-generated layouts to cut out parts.

Manual nesting of parts may require several tries at laying out the parts to achieve the least possible scrap.

Figure 10-11 Using a sqare to draw a straight line.

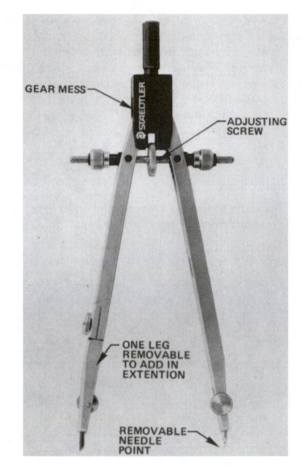

Figure 10-12 Protractor.

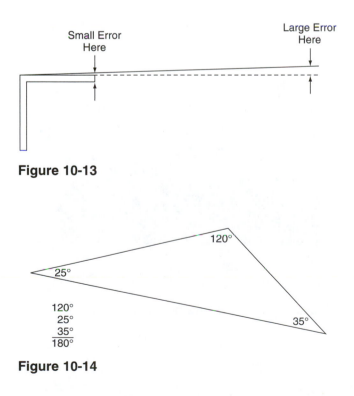

Figure 10-13

Figure 10-14

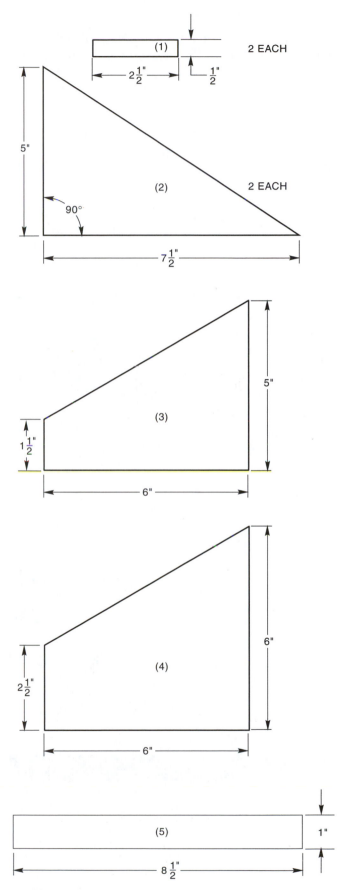

(1)

2½"   ½"   2 EACH

5"

(2)   2 EACH

90°

7½"

(3)

5"

1½"

6"

(4)

2½"   6"

6"

(5)   1"

8½"

Figure 10-15 Parts to be nested

Templates and Patterns

Parts can be laid out by tracing an existing part, template, or pattern. If an existing part is used, mark it and be sure to use the same part for each successive layout. Failure to use the same part can result in an ever-increasing size as each part is laid out slightly larger than the previous one.

To ensure that the parts laid out using templates, patterns, or existing parts are as accurate as possible, be sure that the lines you draw are made as close to the edge as possible (Figure 10-16). When these parts are cut out, remove the line with the cut's kerf, leaving as little of the extraned inside edge of the line as possible. This will ensure that the parts will be the correct size once they are cleaned up (Figure 10-17).

A template can be made in the shape of parts to be produced. Templates can be useful if the parts are complex or need to fit existing weldment or if there will be a large number of similar such parts produced. The advantage of using templates is that once they have been produced, they can be used to lay out subsequent parts quickly and accurately. Templates can be made from any sturdy material such as heavy paper, cardboard paper, wood, or sheet metal. The sturdier the material used for the template, the longer it will last.

Special Tools

Specialty tools have been developed to help layout parts; one such tool is a contour marker (Figure 10-18).

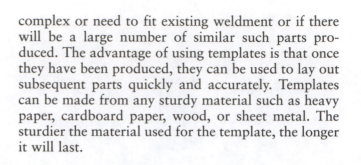

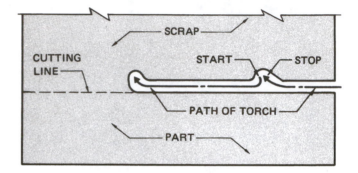

Figure 10-17 Turning out into scrap to make stopping and starting points smoother.

(A)

(B)

Figure 10-16 (A) Tracing a part. (B) Tracing a template.

Figure 10-18 Pipe lateral being laid out with contour marker.

These tools, when used correctly, can help lay out very accurate parts. They require a certain amount of practice to gain experience. Once you have become familiar with the tools, you can lay out an almost infinite variety of joints within the limitations of the tool. An advantage of tools such as the contour marker is that all sides and edges of a structural shape or pipe can be laid out without the necessity of relocating the tool (Figure 10-19).

General Hand Tools

Other useful handheld tools are drills, needle scalers, chipping hammers, chisels, metal shears, nibblers, hack saws, screw drivers, and wrenches of all types and sizes. Hand tools are easily held in the hands when operated. Some power tools are handheld. Many hand tools that can be stored or carried in a mechanic's tool box are useful for some welding-related tasks.

Power Equipment

Larger tools that are frequently used are classified as power tools and equipment. These items range from tools as small as bench grinders and portable equipment such as floor jacks that can be moved in and around the shop to large stationary equipment weighing several tons. Larger power tools that are not small enough to hold are considered power tools and equipment.

Common equipment found in a small welding shop includes tools and equipment used for cutting, drilling, grinding, or sawing metal of various thicknesses. Tools such as the abrasive chop saw, bench grinder, pedestal grinder, drill press, band saw, press breaks, metal rollers, metal benders, and metal shears are used in most welding shops. A machine that combines many of these functions is called an *ironworker*.

FABRICATION

Fabrication is the process of bringing together various pieces, placing them in the correct locations, and welding them to form a completed weldment. A *weldment* refers to an assembly of parts that has been welded together. A weldment may form a completed project itself, or it may be only a single part of a much larger structure. A weldment may consist of two or three pieces such as those used to form a bracket, or it may consist of hundreds or even thousands of individual parts that have been assembled into a much larger structure (Figure 10-20). Even the largest welded structures start by placing two parts together.

All welding projects start with a plan that can range from a simple one that exists only in the mind of the welder to one that has been fully engineered and specified through a complete set of plans, including drawings. Drawings may range from a few lines sketched on a note pad to those created by a computerized drafting program. Throughout their careers, welders will have an opportunity to work with every level, type, and style of weldment plan.

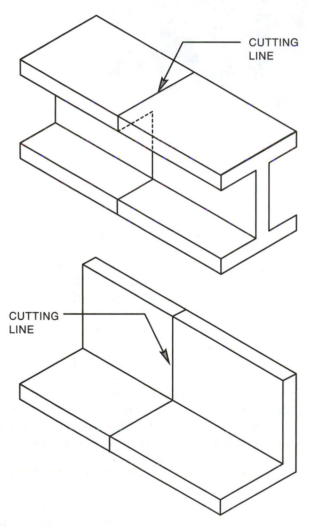

Figure 10-19

Figure 10-20 Large welded oil platform. (Courtesy of Amoco Corporation)

ASSEMBLY

The assembly process consists of bringing together all of the various components and placing them securely in their proper locations using clamps, fixtures, or tack welds in preparation for welding. Assembly requires your proficiency in several areas. You must be able to read specifications and interpret drawings in order to properly locate and assemble all of the individual parts of a project.

Assembly drawings show—both graphically and dimensionally—all of the information that is required to properly locate the various parts of the weldment. These drawings may include either pictorial or exploded views that make it much easier for the beginning assembler to visualize the completed project. Most assembly drawings, however, contain only two, three, or more orthographic views of the project, thus requiring a higher level of understanding and visualization by the assembler (Figure 10-21).

On some projects you begin the assembly by placing parts together along their edges (Figure 10-22). On larger, more complex parts, a center line or baseline may be drawn from which all other dimension locations are measured or taken (Figure 10-23). On some projects, such as this bearing bracket, dimensions between two points are more critical than others. The assembler will locate other parts within tolerance so that the more critical dimension is maintained (Figure 10-24).

The order of assembling parts may be specified on the drawings so that tolerance and accessibility to make the necessary welds does not become a problem. At other times the sequence of assembly is strictly up to the individual worker. Often there is a single part that may be designated as the base, and all other parts will be joined to this central piece.

When starting an assembly, select the largest or most central part, which will be the base for your assembly. Once this is done, all of the other parts can be aligned to this one. Using a base component from which all other dimensioning is done helps to reduce location or dimensioning errors. A slight misalignment of one part, even within tolerances, can become compounded if another part is dimensioned from the misaligned part (Figure 10-25). Thus, using a baseline or center line will result in a more accurate overall assembly.

It is important that you identify each of the component parts and mark them before beginning the actual assembly. It may be helpful for you to hold the parts together and orient them in the same

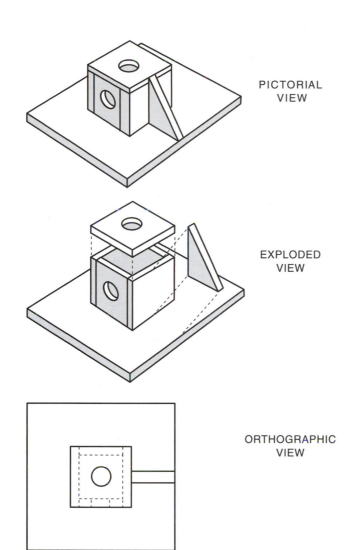

PICTORIAL
VIEW

EXPLODED
VIEW

ORTHOGRAPHIC
VIEW

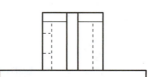

Figure 10-21

Figure 10-22

Figure 10-23

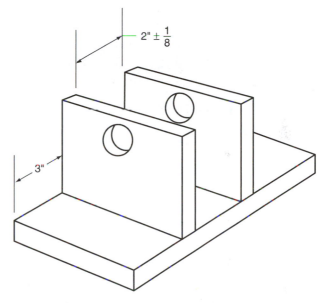

$2'' \pm \frac{1}{8}$

$3''$

Figure 10-24

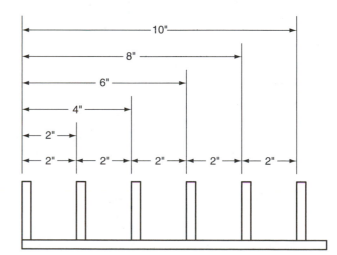

10"

8"

6"

4"

2"

2" 2" 2" 2" 2"

Figure 10-25

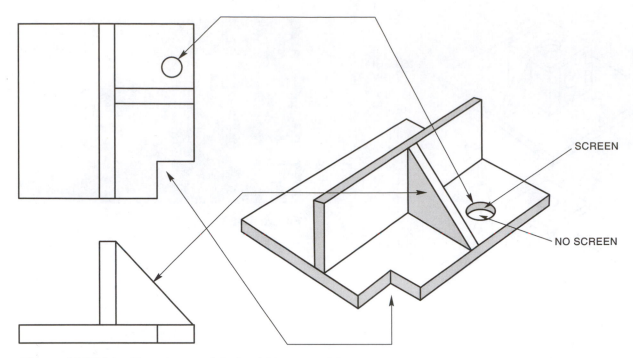

Figure 10-26 Identify unique points to aid in assembly.

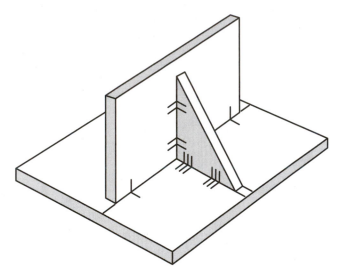

Figure 10-27 Lay out markings to help locate the parts for tack welding.

direction as they are depicted in the drawings. Some parts are more easily identified than others because of unique features such as notches, holes, or unusual shapes that can both help to identify them and aid in locating them (Figure 10-26).

Lines may be drawn on the base to aid in the location of other parts. When a single line is used for this purpose, it is often a good idea to place an X on the side of the line that the part will be placed on. Another method is to draw parallel lines between which the part will be located (Figure 10-27). If two parts are being temporarily located for later assembly, you may want to draw small parallel location marks on both parts where they meet to aid in their relocation.

After the parts have been identified, marked, and located, they are ready for assembly. Parts can be held or clamped into place in preparation for tack welding. Holding parts in place by hand for tack welding is faster than clamping but often leads to errors as the parts may slip while the welder is preparing to begin the tack weld. The more accurate the assembly requirements, the more critical is the clamping. In addition, gas tungsten arc welding often requires the use of both hands to make tack welds.

ASSEMBLY TOOLS

Clamps

A variety of clamps can be used to temporarily hold parts in place so that they can be tack welded.

- *C-clamps* are one of the most commonly used clamps and come in a variety of sizes (Figure 10-28). Some C-clamps have been specially designed for welding. Some have a spatter cover over the screw, and others have their screws made of spatter-resistant materials such as copper alloys.

- *Bar clamps* are useful for clamping larger parts. Bar clamps have a sliding lower jaw that can

Figure 10-28 C-clamps. (Courtesy of Stanley-Proto Industrial Tools, Covington, GA)

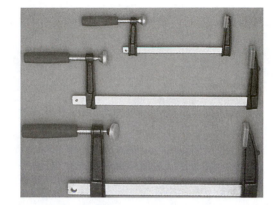

Figure 10-29 Bar clamps. (Courtesy of Woodworker's Supply Inc.)

SLIP JOINT

NEEDLE OR LONG NOSE

LOCKING

Figure 10-30 Three common types of pliers. (Courtesy of Stanley-Proto Industrial Tools, Covington, GA)

be snugged up against the part before tightening the screw clamping end, Figure 10-29. They are available in a variety of lengths.

- *Pipe clamps* are very similar to bar clamps. The advantage of pipe clamps is that the ends can be attached to a section of standard 1/2 in. pipe. This allows for greater flexibility in length, and the pipe can easily be changed if it becomes damaged.

- *Locking pliers* are available in a range of sizes with a number of various jaw designs, Figure 10-30. The versatility and gripping strength make locking pliers very useful. Some locking pliers have a self-adjusting feature that allows them to be moved between different thicknesses without the need to readjust them.

- *Cam-lock clamps* are specialty clamps that are often used in conjunction to a jig or fixture. They can be preset which allows for faster work, Figure 10-31.

- *Specialty clamps* such as those in Figure 10-32 are for pipe welding and many other different types of jobs. Such specialty clamps make it possible to do faster and more accurate assembling.

Fixtures

Fixtures are devices that are made to aid in the assembly and fabrication of weldments. Because of the time and expense associated with the fabrication of fixtures, they are typically used only when a large number of similar parts are to be fabricated or when their use is required for accurate assembly. When used, they must be strong enough to support both the weights of the parts and weld stresses experienced during assembly and remain in tolerance. Some fixtures have manual or automatic clamping devices that aid in their use. Locating pins or other devices can ensure proper part location. A properly designed weld fixture allows adequate room for the welder to easily perform the necessary tack welding. Some

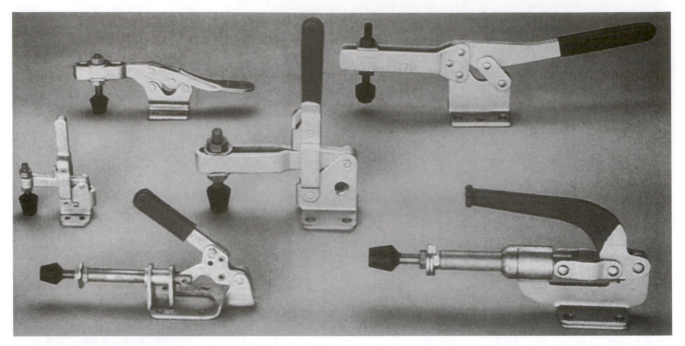

Figure 10-31 Toggle clamps. (Courtesy of Woodworker's Supply Inc.)

(A)

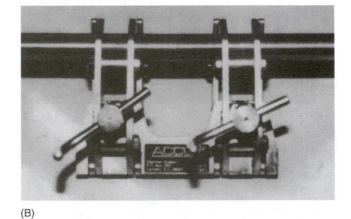

(B)

Figure 10-32 (A) Pipe alignment clamps. (B) Pipe clamps.

parts may be left in the fixture throughout the entire welding process in order to reduce distortion.

Welding fixtures allow you to speed up the operation of hand fitting and squaring welded parts. Even with very little fit-up experience, you can use fixtures to produce assemblies that look good and are well made.

PLATE PREPARATION

Before welding on carbon steel, you should remove any surface material such as dirt, oil, paint, grease, or other contaminants from the metal.

Even if the carbon steel material looks clean, it could be covered with oxides, which are a protective coating caused by moisture in the atmosphere. Moisture adheres to the base metal, slightly oxidizing it to form a gray color that covers new or cleaned material after a few days' exposure. If left exposed for long periods of time, this protective coating corrodes the metal. The color darkens, eventually turning to rust. This condition is known as oxidation, or the wasting away of metal, and the metal is covered with iron oxide. Metal in this condition is commonly called *rusty*. If the oxidization continues, the metal corrodes to an unusable state and eventually crumbles into grit.

We want to prevent this corrosion because we do not want to weld over a rusty surface. Surface contamination with oxides causes impurities in the base metal.

CAUTION

Always wear safety glasses when cleaning dirty or rusty materials.

METAL CLEANING TOOLS

Brushes

The most practical and least expensive metal cleaning tool is a handheld wire brush. Wire brushes range from less than an inch to several inches in size and come in many shapes. Some are equipped with a combination chipping tool and/or built-in scraper. Wire brushes are available in different materials from stainless steel to brass bristles.

Abrasive Grinding

Preparing weld joints by using an abrasive grinding process produces no mill scale and cleans the surface to a bright metallic sheen that is ideal for welding. Grinding is noisy and produces lots of flying sparks and grit. It is generally accomplished with a handheld power tool, so its results depend on the operator's skill.

Power Tools

Other tools for removing surface oxides are handheld power grinders, sanders, wire wheels, and bench pedestal grinders. These tools can be electric-powered tools or air-powered, pneumatic tools.

CAUTION

Remove oxides from the weld joint or base metal surfaces. Otherwise, the weld may not fuse properly, or inclusions may be trapped in the weld area.

FITTING

Not all parts easily fit together exactly as they were designed. There may be slight imperfections such as those caused by cutting errors or distortion that may prevent them from fitting into their exact location. Occasionally such problems can be resolved by grinding or filing the part to fit. Hand grinders can be used for such fitting; however, you must be careful not to remove excessive material (Figure 10-33). Some metals such as aluminum require special grinding stones. On small or delicate components such as those fabricated from sheet metal, filing the parts to fit may be the only solution. Rubbing the filed face with a soapstone before it is used can help prevent metal chips, such as those from soft metal such as aluminum and brass, from becoming stuck in the file's surface. Some fitting requires that the parts be forced into alignment. One way to do this is to make a small tack weld in the joint and then use a hammer and anvil to pound the parts into alignment (Figure 10-34). Small tack weld used in this manner will become part of the finished weld.

TACK WELDING

Tack welds are small welds that are made to hold parts in place until they can be welded. Tack welds must be made in such a manner that they can be easily incorporated into the finished weld. They should be made using the same welding procedure and filler metal that will be used for the finished

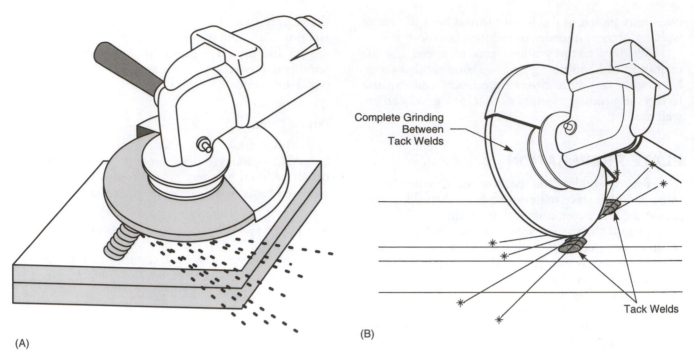

(A)

(B)

Figure 10-33 (A) (Courtesy of Mike Gellerman) (B) (Courtesy of Mike Gellerman)

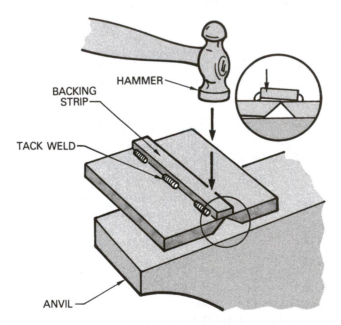

Figure 10-34 Using a hammer to align the backing strip and weld plates.

weld. Tack welds must be large enough to withstand the forces from assembly and welding yet small enough to be easily incorporated into the finished weld without causing a discontinuity in its size or shape (Figure 10-35).

The size and location of a tack weld will vary from weldment to weldment. Some thick sections may require a tack weld to be made only at each end of the joint; however, on some thin sections where distortion can become a problem, tack welds may need to be

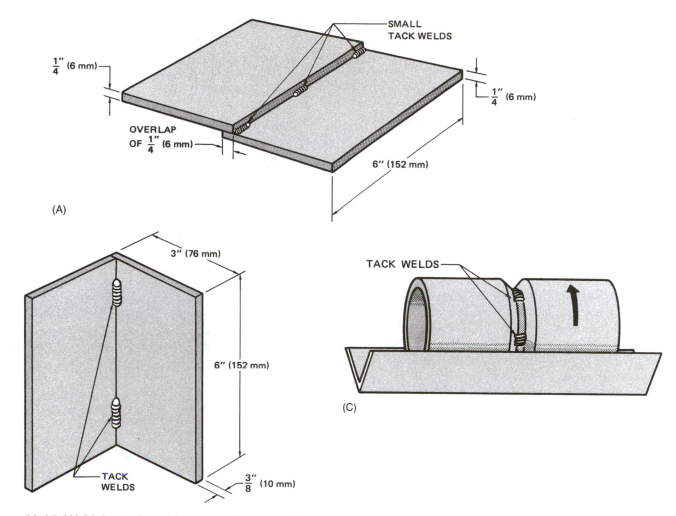

Figure 10-35 (A) Make tack welds as small as possible.

made closer together. Do not hesitate to add additional tack welds to a joint if, during the welding process, you notice that they are needed to prevent distortion.

WELDING

The order and direction of welds can significantly affect weldment distortion. Generally welds should be staggered from one location on a part to another. This allows the welding heat and its related stresses to dissipate so that they do not result in distortions.

When starting a weld, be sure that the arc is struck only in the joint where the weld will be made. Arc strikes outside the joint can result in unsightly marks that must be removed following welding and, under some code conditions, are considered to be welding defects.

Before you begin welding, check your range and freedom of movement, especially for gas tungsten arc welding, because you must maintain a very accurate arc length throughout the weld. Any interference with the manipulation of the filler metal and torch can easily disrupt this arc distance. If the tungsten is accidentally dipped into the molten weld pool, significant clean-up time may result.

New welders tend to place the practice plate squarely on the table; however, this may not be the most ideal position for you. With the power off and your welding hood up, make circle practice runs along the joint with the plate at various angles to your body (Figure 10-36). You may find that one angle enables you to follow the joint more easily than others.

As you develop your GMA welding skills, practice making welds in different positions. Not many field welds are produced on parts that are held in

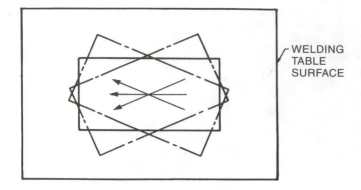

Figure 10-36 Change the plate angle to find the most comfortable welding position.

your most ideal welding position. Improving your skills in these out-of-position stances will be beneficial to you.

If you are working on a weldment that is too large to fit into a welding booth, use portable welding screens to protect others in the welding area from the welding arc's light. Be sure to follow all safety procedures recommended by your welding shop.

REPAIR WELDING

GMAW is used to repair a wide variety of metal components. Some welding shops may specialize in welding repair work. These shops may even contract with large production facilities to do their repair work. Large fabrication shops often have a product line and specialize in only one area of welding. Their tools are generally set up for specific production, and they rarely do welding repairs.

GMAW is a good way to repair metal items of all types. Apply the knowledge and experience gained through the practice exercises in this book, and you can fix things as good as new or better.

Repair welding usually, but not always, means fixing something that is broken. The natural elements sometimes affect the integrity, character, or construction of a material by causing rust. Local agencies salt streets and highways to keep them safer for drivers. This salt is a corrosive chemical and causes steel to rust rapidly. Rusted areas can be easily replaced by welding, or if rusting is not too advanced, base metal may be resurfaced, similar to the process used to make practice padding plates.

Preparation for Repair Welding

Have all safety equipment easily accessible and in working condition for repair welding. Even if it is a small project, do not risk injury. Use personal protective equipment.

Organize all the necessary tools. A plan, shop drawing, or a print is always helpful. Plan the welding repair before starting. Not all repairs need a complicated plan. Just use common sense and welding knowledge to determine the best way to do the job.

For best results use a grinder or wire brush to clean all metal parts. Use clamps to secure and square parts in a fixture, and check them with a squaring tool. Make tack welds and recheck the parts for square before welding.

CAUTION

Always remove combustibles from the area before striking a welding arc.

General GMA Welding Repairs

Welding repairs may include items from the home, office, manufacturing, recreation, transportation, or virtually any area. You may be asked to repair a kitchen chair, tricycle, basketball hoop and backboard, steel patio table, truck frame, trailer hitch, boat, and many other items. Here are some important things to remember when repairing these items:

1. Follow all safety precautions.
2. Clean the surface of the repair area.
3. Align the parts carefully.
4. Use clamps and fixtures to position and hold parts throughout the welding process to prevent warpage and distortion.
5. Bevel the weld joint areas if necessary.
6. Use the correct welding parameters, electricity, and shielding gas.
7. Never weld on concrete floors because moisture trapped in the concrete may expand and the concrete may explode.
8. Use care when grinding, cleaning, and painting to produce a safe and quality repair.

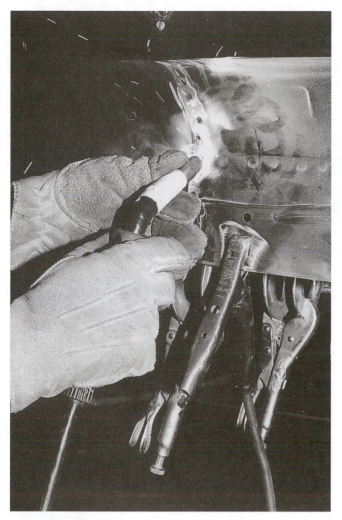

Figure 10-37

Automobile Repair Welding

To protect it, always disconnect the auto's electrical system before welding. Remove the negative (−) cable and/or the chaise ground wire from the vehicle's battery terminal post. It is important that you do not allow electrical auto components to lie between the weld zone and the workpiece connection (Figure 10-37). If the auto components can be removed before welding, do so.

CAUTION

Always post a fire watcher with the appropriate type of fire extinguisher when welding on autos.

When welding on autos, make sure the work area is free of flammable material and a fire extinguisher is within reach. Using appropriate covers, protect interior and exterior areas, including window glass, of the auto from welding sparks. The glass may pit or exterior paint may be burned or discolored from weld spatter and heat.

Some newer vehicles have a supplemental inflatable restraint (SIR) or air bag. Disconnect the unit to prevent discharge. Refer to the auto manufacturer's repair manual or the service manual for location of and procedures for handling the SIR before attempting welding repairs.

CAUTION

Failure to disable the SIR may result in air-bag deployment. Injury to the welder or helpers and/or damage to the SIR system may occur.

FINISHING

Depending on the size of the shop, the welder may be responsible for some or all of the finish work. This may include chipping, cleaning, or grinding the welds to applying paint or other protective surfaces.

Grinding of welds should be avoided if possible by properly sizing the weld as it is made. Grinding can be an expensive process, adding significant cost to the finished weldment. Sometimes it is necessary to grind for fitting purposes or for appearance, but even in these cases it should be minimized.

CAUTION

When using a portable grinder, be sure that it is properly grounded, the sparks will not fall on others, cause damage, or start a fire, and maintain control to prevent the stone from catching and gouging the part or yourself.

CAUTION

Be sure that any stone or sandpaper is rated for a speed in revolutions per minute (rpm) that is equal to or greater than the speed of the grinder itself. Using a stone with a lower-rated rpm may result in the stone's flying apart with explosive force.

Figure 10-38 Wire brushes and grinding stones used to clean up welds.

Most grinding is done with a hand angle grinder (Figure 10-38). These grinders can be used with a flat or cupped grinding stone or sandpaper. As the grinder is used, the stone will wear down. Once the grinding stone has worn down to the paper backing, it must be discarded. It is a good practice to hold the grinder at an angle so that, if anything is thrown off of the stone or metal surface, it will strike no one in the area. Because of the speed at which the grinding stone turns, any such object can cause serious injury.

The grinder must be held securely so that there is a constant pressure on the work. If the pressure is too great, the grinder motor will overheat and may burn out. If the pressure is too light, the grinder may bounce and thus crack the grinding stone. Move the grinder in a smooth pattern along the weld. Watch the weld surface as it begins to take the desired shape, and change your pattern as needed.

Painting and other finishes release fumes such as volatile organic compounds (voc) that are often regulated by local, state, and national governments. Special ventilation is required for most paints. Such a ventilation system will remove harmful fumes from the air before it is released back into the environment. Check with your local, state, or national regulating authority before using such products. Read and follow all manufacturer's instructions for the safe use of their product.

CAUTION

Most paints are flammable and must be stored well away from any welding.

SUMMARY

For most welders the feeling you get from building or repairing something yourself is extremely rewarding. The pride and satisfaction from this type of work makes the long hours required to learn to become a good welder worthwhile. Part of this satisfaction comes from the fact that layout, fabrication, and repair work demonstrates that you have mastered not only welding but also a number of other skills. Improving those skills—like welding—will come from practice. At any opportunity, even with parts tacked together for lab practice, do your best to make sure that they are square and properly sized and have a quality appearance.

REVIEW QUESTIONS

1. Define *layout*.
2. The process of laying out a part may be affected by what factors?
3. What is used to lay out circles, arcs, and curves?
4. What must the interior angles of all triangles add up to?
5. Define *nesting*.
6. What is the advantage of using templates?
7. What is one example of a specialty tool developed to aid in the laying out of parts?
8. Name three other general handheld tools that are useful?
9. Larger power tools that are not small enough to hold are considered what?
10. What does the term *weldment* refer to?
11. What is an assembly drawing?
12. What are the most commonly used clamps?
13. Fixtures are devices that are made up to aid in _____.
14. What is the most practical metal cleaning tool?
15. What is a tack weld?
16. What should be used to figure out the best way to do a repair welding job?
17. What should always be disconnected when doing automotive repair?
18. _____ should be avoided if possible by properly sizing the weld as it is made.

CHAPTER 11

QUALIFIED AND CERTIFIED WELDERS

Welding, in most cases, is one of the few professions that requires job applicants to demonstrate their skills even if they are already certified. Other professions (for example, doctor, lawyer, and pilot) may require a written test or license. Welders, however, are often required to demonstrate both their knowledge and skills before they are hired since welding, unlike most other occupations, requires a high degree of eye–hand coordination.

A method commonly used for welders to demonstrate their welding ability is for them to take a qualification or certifications test. Welders who have passed such a test are referred to as qualified welders, and if the proper written records are kept as proof of the test results, then they are referred to as certified welders. The basic difference between a qualified welder and a certified welder is that written records are kept for certified welders (Figure 11-1). Not all welding jobs require that a welder be certified. Some jobs require only that welders pass a basic weld test before they are hired.

Welder certification is divided into two major areas. The first area covers the traditional welder certification that has been used for years. This type of certification would be used when a welder takes a welding test to demonstrate the welding skills required for a specific process on a specific weld. This test may be taken to qualify for a welding assignment or to meet the requirements for employment.

The second, and newer, area of certification has been developed by the American Welding Society.

This certification has three levels. The first level is primarily designed for a new welder who needs to demonstrate Entry-Level Welder's skills. The other levels cover Advanced Welders and Expert Welders. This chapter covers the traditional certification and the AWS QC10 Specification for Qualification and Certification for Entry-Level Welder.

CODE OR STANDARDS REQUIREMENTS

The type, depth, angle, and location of a groove is usually determined by a code or standard that has been qualified for the specific jobs. Organizations such as the American Welding Society, American Society of Mechanical Engineers, and the American Bureau of Ships issue such codes and specifications. The most common code or standards are the AWS D1.1 and the ASME Boiler and Pressure Vessel (BPV) Section IX.

Some joint designs that have been tested and found to be reliable for the weldments for specific applications are called *prequalified*. Joint designs that have been modified and are no longer considered prequalified must be tested to make certain that they will be reliable.

QUALIFIED AND CERTIFIED WELDERS

Welder qualification and welder certification are often misunderstood. Sometimes it is assumed that a qualified or certified welder can weld anything.

WELDER AND WELDING OPERATOR QUALIFICATION TEST RECORD (WQR)

Welder or welding operator's name _____ Identification no. _____

Welding process _____ Manual _____ Semiautomatic _____ Machine _____

Position _____

(Flat, horizontal, overhead, or vertical - if vertical, state whether up or down.)
In accordance with welding procedure specification no._____

Material specification _____

Diameter and wall thickness (if pipe) - otherwise, joint thickness _____

Thickness range this qualifies_____

FILLER METAL

Specification No. _____ Classification _____ F-number _____

Describe filler metal (if not covered by AWS specification) _____

Is backing strip used? _____

Filler metal diameter and trade name _____ Flux for submerged arc or gas
for gas metal arc or flux-cored arc welding _____

GUIDED BEND TEST RESULTS

Appearance _____ Weld Size _____

Type   Result   Type   Result

Test conducted by _____ Laboratory test no. _____

FILLET TEST RESULTS

Appearance _____ Fillet Size _____

Fracture test root penetration _____ Macroetch _____

(Describe the location, nature, and size of any crack or tearing of the specimen.)
Test conducted by _____ Laboratory test no._____

RADIOGRAPHIC TEST RESULTS

FILM IDENTIFICATION	RESULTS	REMARKS	FILM IDENTIFICATION	RESULTS	REMARKS

Test conducted by _____ Laboratory test no._____

   We the undersigned, certifiy that the statements in this record are correct and that the welds were prepared and tested in accordance to these requriements.

Manufacturer or contractor _____

Authorized by _____

Date _____

Figure 11-1

Being certified does not mean that a welder can weld everything, nor does it mean that every weld that is made is acceptable. It means that the welder has demonstrated the skills and knowledge to make good welds. To ensure that a welder is consistently making welds that meet the standard, welds are inspected and tested. The more critical the welding, the more critical the inspection and the more extensive the testing of the welds.

All welding processes can be tested for qualification and certification. The testing can range from making spot welds with an electric resistant spot welder to making gas tungsten arc electron beam welds on aircraft. Being qualified or certified in one area of welding does not automatically mean that you can make quality welds in other areas. Most qualification and certifications are restricted to a single welding process, position, metal, and thickness range.

Changes in any one of a number of essential variables can result in the need to recertify (Figure 11-2).

- Process–Welders can be certified in each welding process such as SMAW, GMAW, FCAW, GTAW, EBW, and RSW. Therefore, a new test is required for each process.

- Material–The type of metal, such as steel, aluminum, stainless, and titanium, being welded will require a change in the certification. Even a change in the alloy within a base metal type can require a change in certification.

PROCEDURE QUALIFICATION RECORD (PQR)

Welding Qualification Record No: _____ WPS No: _____ Date: _____

Material specification _____ to _____

P-No. _____ Grade No. _____ to P-No. _____ Grade No. _____ Thickness and O.D. _____

Welding process: Manual _____ Automatic _____

Thickness Range _____

FILLER METAL

Specification No. _____ Classification _____ F-number _____

A-number _____ Filler Metal Size _____ Trade Name _____

Describe filler metal (if not covered by AWS specification) _____

_____

FLUX OR ATMOSPHERE

Shielding Gas _____ Flow Rate _____ Purge _____

Flux Classification _____ Trade Name _____

WELDING VARIABLES

Joint Type _____   Position _____

Backing_____   Preheat _____

Passes and Size _____   Bead Type _____

No. of Arcs _____   Current _____

Ampere_____   Volts _____

Travel Speed_____   Oscillation _____

Interpass Temperature Range _____

WELD RESULTS

Appearance _____   Weld Size _____

Figure 11-2

GUIDED BEND TEST

TYPE	RESULT	TYPE	RESULT

TENSILE TEST

SPECIMEN NO.	DIMENSIONS WIDTH/THICKNESS	AREA	ULTIMATE TOTAL LOAD, LB.	ULTIMATE UNIT STRESS, PSI.	CHARACTER OF FAILURE AND LOCATION

Welder's Name _____ Identification No. _____ Laboratory Test No. _____

By virtue of these tests meets welder performance requirements.

Test Conducted by _____ Address _____

per _____ Date _____

We certify that the statements in this record are correct and that the test welds were performed and tested in accordance to the WPS.

Manufacturer _____

Signed by _____

Date _____

Figure 11-2 continued

- Thickness–Each certification is valid on a specific range of thickness of base metal. This range depends on the thickness of the metal used in the test. For example, if a 3/8″ (9.5 mm) plain carbon steel plate is used, then under some codes the welder would be qualified to make welds in plate-thickness ranges from 3/16″ to 3/4″ (4.7 mm to 19 mm).

- Filler metal–Changes in the classification and size of the filler metal can require recertification.

- Shielding gas–If the process requires a shielding gas, then changes in gas type or mixture can affect the certification.

- Position–In most cases, if the weld test was taken in the flat, the certification is limited to flat and possibly horizontal. If the test was taken in the vertical, the welder is usually allowed to work in the flat, horizontal, and vertical positions.

- Joint design–Changes in weld type, such as groove or fillet welds, will require a new certification. Additionally, variations in the joint geometry, such as groove type, groove angle, and number of passes, can also require retesting.

- Welding current–In some cases changing from AC to DC or changes such as pulsed power and high frequency can affect the certification.

Any welder qualification or certification process must include the specific welding skill level that is to be demonstrated. The detailed information for welding and testing is often given as part of a Welding Procedure Specification (WPS) or similar set of welding specification or schedule (see Chapter 13 and Figure 11-3). Such a specific set of written standards is needed so that everyone knows what skills are required. The WPS enables both the welder to be prepared for the required welding test and the company

WELDING PROCEDURE SPECIFICATION (WPS)

Welding Procedures Specifications No: _____ Date: _____

TITLE:

Welding _____ of _____ to _____

SCOPE:

This procedure is applicable for _____

within the range of _____ through _____

Welding may be performed in the following positions _____

BASE METAL:

The base metal shall conform to _____

Backing material specification _____

FILLER METAL:

The filler metal shall conform to AWS specification No. _____ from AWS

specification _____. This filler metal falls into F-number _____

and A-number _____

SHIELDING GAS:

The shielding gas, or gases, shall conform to the following compositions and purity:

_____

JOINT DESIGN AND TOLERANCES:

PREPARATION OF BASE METAL:

ELECTRICAL CHARACTERISTICS:

The current shall be _____

The base metal shall be on the _____ side of the line.

PREHEAT:

BACKING GAS:

WELDING TECHNIQUE:

INTERPASS TEMPERATURE:

CLEANING:

INSPECTION:

REPAIR:

SKETCHES:

Figure 11-3 Sample Welding Procedure Specification (WPS) Form.

to know what skills the welder has demonstrated. Varying from these strict limitations in the WPS usually requires that a different test be taken.

Welder Performance Qualification is the demonstration of a welder's ability to produce welds that meet very specific prescribed standards. The form used to document this test is called the Welding Qualification Test Record. The detailed written instructions that are to be followed by the welder are called the Welder Qualification Procedure. Welders who pass this are often referred to as being Qualified Welders or just Qualified.

Welder Certification is the written verification that a welder has produced welds that meet a prescribed standard of welder performance. A welder who holds such a written verification is often referred to as Certified or a Certified Welder.

An AWS Certified Welder is one who has complied with all of the provisions, requirements, and specifications of the AWS regarding this certification. Very specific requirements must be followed by any school or organization before they can offer this certification. Under the AWS program the welder must pass a closed-book exam over specific knowledge areas and a performance test. Written documentation including the welder's name and Social Security number along with test results must be sent to the AWS.

These records are entered into the AWS National Registry for welders. The certification record expires after one year and at that time are automatically deleted from the registry.

REVIEW QUESTIONS

1. What is a common method used for welders to demonstrate their welding ability?
2. What is the basic difference between a qualified welder and a certified welder?
3. What are the three levels of the newer area of certification that has been developed by the AWS?
4. What are the type, depth, angle, and location of a groove usually determined by?
5. What are the most common codes or standards?
6. What does being *certified* mean?
7. What are most qualifications and certifications restricted to?
8. Changes in the _____ and _____ of the filler metal can require recertification.
9. What is Welding Performance Qualification a demonstration of?
10. What is an AWS certified welder?

INSPECTION REPORT		JOB: _____	
INSPECTION	PASS/FAIL	INSPECTOR'S INITIALS	DATE
Layout			
Cutout			
Assembly			
Welding			
Tack			
Interpass			
Finish			
Overall Rating			
Accuracy			
Appearance			

Welder _____ Date _____

(B)

Figure 12-1 continued

BILL OF MATERIALS		
Name _____ Date _____ Job _____		
Part ID	Size Determination	SI Determination

Material Specification _____

(C)

Figure 12-1 continued

example of the type of form that may used for keeping and maintaining these records.

Verbal Instructions

From time to time on any job, additional instructions must be given to the welder verbally. These instructions are as important as those provided in writing. Many times in a welding shop minor modifications of a weld may be required for meeting a specific customer's requirements or enabling the shop to take advantage of existing materials. In some cases critical information such as particular safety concerns may be given verbally. It is your responsibility to remember and follow all such verbal instructions.

Safety

Job safety is a major concern for everyone. Welding can present specific concerns that must be addressed by everyone working in a shop. Specific safety information pertaining to gas metal arc welding is included in Chapter 2, and additional information is available in *Safety for Welders* by Larry Jeffus and ANSI Z49.1. Safety is such an important area of concern that a score of 100% is required before you are allowed to do any work in the welding lab.

Housekeeping

All welding processes produce some level of scrap or trash. Additionally the work area can become cluttered with welding leads, hoses, torches, hand tools, power tools, clamps, jigs, fixtures, and other items. It is the responsibility of the welders to maintain their work space in a clean and an orderly manner.

On some welding jobs, some degree of housekeeping may be provided; however, welders are usually held responsible for their specific work station's condition. Learning good housekeeping skills, therefore, is important.

Routine repairs of equipment may also fall under housekeeping chores. Most welders are expected to change their own welding hood lenses as needed and to make minor repairs or adjustments to the GMA welding gun and work plant. Before making any such repairs or adjustments, you must first study the manufacturer's literature for the equipment, check with your instructor, and turn off the power.

Cutting Out Parts

The parts will be cut out using the manual oxyfuel gas cutting (OFC), machine oxyfuel gas cutting (OFC-track burner), air arc cutting (CAC-A), and manual plasma arc cutting (PAC) processes.

To cut out the various parts, you will need to make straight, angled, and circular cuts. In some cases you may determine which method and process to use. On other cuts the method and process are specified on the drawing. Inspect all cut surfaces for flaws and defects and repair them if necessary (Figure 12-2).

Fit-Up and Assembly of the Parts

Putting the parts of a weldment together in preparation for welding requires special skills. The more complex the weldment, the more difficult the assembling. Each part must be located and squared to other parts (Figure 12-3). To complicate this process, clamping may be necessary because the parts may not all be flat and straight.

As metal is cut, it may become distorted as a result of heat from the cutting process. Any such distortion or bend that may have been in the metal must be corrected during the fit-up process. Sometimes grinding can be required to make a correct fit-up (Figure 12-4). At other times the parts can be fitted together correctly by using C-clamps or pliers (Figure 12-5). In more difficult cases, tack welds and cleats or dogs must be used to achieve the proper fit-up (Figure 12-6).

Workmanship Standards for Welding

The weldment must be positioned so that the welds are made within 5° of the specified position.

All arc strikes must be within the groove. Arc strikes outside of the groove will be considered defects.

Tack welds must be small enough so that they do not interfere with the finished weld. They must, however, be large enough to withstand the shrinkage forces from the welds as they are being made. Sometimes it is a good idea to use several small tacks on the same joint to ensure that the parts are held in place (Figure 12-7).

The weld bead size is important. Beads must be sized in accordance with the WPS or drawing for each specific weld. All weld bead starts and stops must be smooth (Figure 12-8).

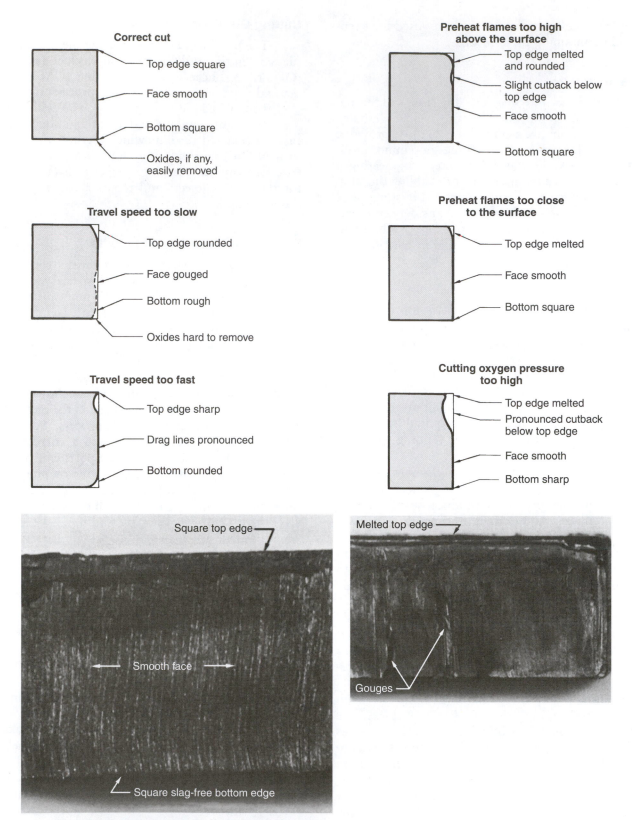

Figure 12-2 Flame-cut profiles and standards.

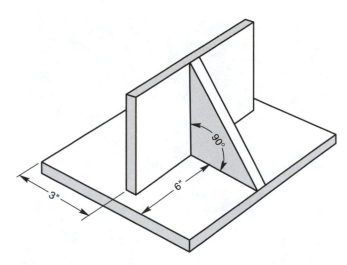

Figure 12-3 Locate and square parts to be welded.

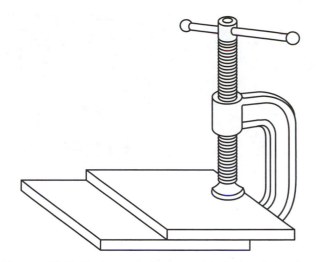

Figure 12-5 C-clamp being used to hold plates for tack welding.

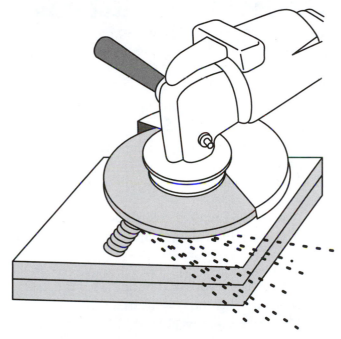

Figure 12-4 A portable grinder can be used to correct cutting or fitting problems.

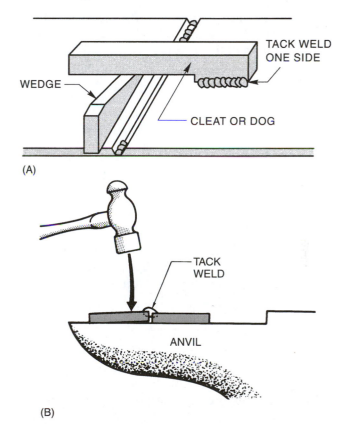

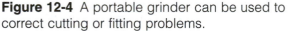

Figure 12-6 (A) Wedges and cleats can be used on heavier metal to pull the joint into alignment. (B) The plates can also be forced into alignment by striking them with a hammer.

The weld beads must be cleaned. All weld cleaning must be performed with the test plate in the welding position. A grinder may not be used to remove weld control problems.

Weld Inspection

Each weld and/or weld pass is inspected visually for the following:

- There shall be no cracks or incomplete fusion.
- There shall be no incomplete joint penetration in groove welds except as permitted for partial joint penetration groove welds.

- The test supervisor shall examine the weld for acceptable appearance and shall be satisfied that the welder is skilled in using the process and procedure specified for the test.
- Undercut shall not exceed the lesser of 10% of the base metal thickness or 1/32″ (0.8 mm).

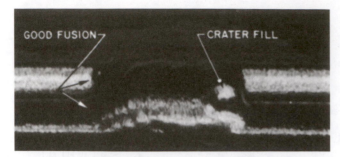

Figure 12-7 Tack weld. Note the good fusion at the start and crater fill at the end.

- Where visual examination is the only criterion for acceptance, all weld passes are subject to visual examination at the direction of the test supervisor.
- The frequency of porosity shall not exceed one in each 4″ (100 mm) of weld length, and the maximum diameter shall not exceed 3/32″ (2.4 mm).
- Welds shall be free from overlap.*

WELDING SKILL DEVELOPMENT

The American Welding Society's *AWS EG2.0-95 Guide for the Training and Qualification of Welding Personnel Entry-Level Welder* lists safety, related knowledge, and welding skills that must be mastered as part of a training program.

The amount of time required for each welder to develop the necessary skills to pass the American Welding Society Certified Welder program will vary greatly from individual to individual.

SUMMARY

Passing the Entry-Level Welder Qualification test will establish your skills on a nationally recognized standard. Earning such recognition will both aid in your job prospects and enhance your self-esteem.

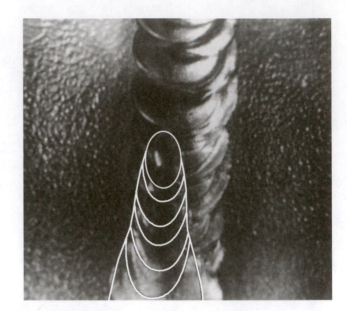

Figure 12-8 Taper down the weld when it is complete or when stopping to reposition. This will make the end smoother and ease restarting the next weld bead if necessary.

REVIEW QUESTIONS

1. What are the two major components of the written exam?
2. What is expected in the layout part of the practical test?
3. When should written records be kept?
4. What score is required on the safety part of the test?
5. Any distortion or bend that may have been in the metal must be corrected during what process?
6. What size must tack welds be?
7. What determines weld bead size?
8. How much time is required for welder to develop the necessary skills to pass the AWS certified welder program?

*American Welding Society AWS QC10-95 Specification for Qualification and Certification for Entry-Level Welders

CHAPTER 13

WELDING SPECIFICATIONS

As a result of years of experience and thousands of tests, specific welding criteria have been established that, when followed, have the greatest likelihood of producing quality welds. These specifications have ranges that allow welders to make changes to suit their individual welding style.

When specific welding standards have been established for a particular weld, this information is provided in a document called welding procedures specifications (WPS). WPS should contain all information required by a welder to produce a satisfactory weld.

WPS GENERAL INFORMATION

A WPS should contain information about the process material, filler metal, shielding gas, and other important items. Figure 13-1 is an example of a blank WPS form that can be used for processes other than gas tungsten arc welding.

1. Number: The WPS number may contain more than simply a numeric identifier, for example, a name (or an abbreviation of a name) that identifies the welding school, shop, or organization that helped develop this specific document.

2. Date: This is the date that the WPS was accepted following testing.

3. Title: The title includes information regarding the welding process that will be used and the types of material that will be joined. The process might be GMAW-S or any other specific process identification.

GAS METAL ARC WELDING-SHORT CIRCUIT METAL TRANSFER (GMAW-S)

WELDING PROCEDURE SPECIFICATION (WPS)
Welding Procedures Specifications No: <u>TEW GMAW 1</u>   Date: _____

TITLE: Welding <u>GMAW-S</u>   of <u>plate</u>   to <u>plate</u>   .

SCOPE: This procedure is applicable for <u>V-groove, bevel and fillet welds</u>   within the range of <u>1/8 in. (3.2 mm)</u>   through <u>1 1/2 in. (38 mm)</u>   . Welding may be performed in the following positions <u>ALL</u>   .

BASE METAL: The base metal shall conform to <u>Carbon Steel M-1, P-1 and S-1 Group 1 or 2</u>
Backing material specification <u>NONE</u>   .

FILLER METAL: The filler metal shall conform to AWS specification <u>No. E70S-3</u>   from AWS specification <u>A5.18</u>   . This filler metal falls into F-number <u>F-6</u>   and A-number <u>A-1</u>   .

SHIELDING GAS: The shielding gas, or gases, shall conform to the following compositions and purity:
CO_2 at 30 to 50 CFH or 75% Ar/CO_2 25% at 30 to 50 CFH.

Figure 13-1

Items 3b and 3c are for the types of material, sheet, plate, pipe, or tubing that are to be welded. On some WPS forms, plate-to-pipe welds, plate-to-tubing welds, tubing welds, or any other such combination may be included here.

4. Scope: Included in 4a is the type of weld (for example, a groove weld, a fillet weld, or a plug weld).

   Items 4b and 4c indicate the thickness range. This might be for a single range (such as 1/4″ to 1/4″) or for a more inclusive range (such as 18 gauge to 10 gauge).

   Item 4d lists the welding positions that this WPS covers. Single or multiple positions can be included such as 1G and 2G or 1G, 2G, 3G, and so on.

5. Base metal: Specific information regarding the makeup of the base material is included. General information such as mild steel, stainless steel, aluminum, and titanium is included with the specific grouping of the base materials.

   Item 5b includes specifications for any backing material that was used for the joint such as backing strips that might be used for a V-groove butt joint.

6. Item 6a includes specific information regarding the American Welding Society's identification number for the filler metal.

   Item 6b is the specification documentation under which the specific filler metal in 6a can be found.

   Item 6c is the F-number group designation for the metal type (Table 13-1).

   Item 6d is the A number, which is an additional identification number used for some metals (Table 13-2).

7. Shielding gas: The specific makeup of the shielding gas (whether it is 100% CO_2, CO_2 argon, argon–oxygen, or any other gas or mixture of gases) is included with the grade, which is usually denoted as welding grade.

8. Joint design and tolerance: A cross-section of the specific welding joint such as the one illustrated in Figure 13-2 is included. This drawing should include acceptable limits on groove and joint angles as well as root spacing, backing material, sizes, and so on.

9. Preparation of base metal: Specific information regarding cleaning and preparation of the base metal should be included. In most cases, for code-quality welds the joint must be cleaned thoroughly around a minimum of 1 inch of the groove on both the inside and outside surfaces. Such details should be included.

TABLE 13-1

F NUMBERS		
GROUP DESIGNATION	**METAL TYPES**	**AWS ELECTRODE CLASSIFICATION**
F1	Carbon Steel	EXX20, EXX24, EXX27, EXX28
F2	Carbon Steel	EXX12, EXX13, EXX14
F3	Carbon Steel	EXX10, EXX11
F4	Carbon Steel	EXX15, EXX16, EXX18
F5	Stainless Steel	EXXX15, EXXX16
F6	Stainless Steel	ERXXX
F22	Aluminum	ERXXXX

TABLE 13-2

SPECIFICATION NUMBERS	
A5.10	Aluminum—bare electrodes and rods
A5.3	Aluminum—covered electrodes
A5.8	Brazing filler metal
A5.1	Steel, carbon, covered electrodes
A5.20	Steel, carbon, flux-cored electrodes
A5.17	Steel-carbon, submerged arc wires and fluxes
A5.18	Steel-carbon, gas metal arc electrodes
A5.2	Steel—oxyfuel gas welding
A5.5	Steel—low-alloy covered electrodes
A5.23	Steel—low-alloy electrodes and fluxes—submerged arc
A5.28	Steel—low-alloy filler metals for gas shielded arc welding
A5.29	Steel—low-alloy, flux-cored electrodes

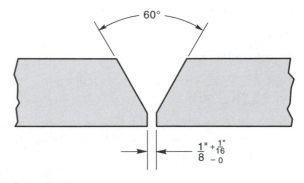

Figure 13-2

TABLE 13-3

ELECTRODE		WELDING POWER			SHIELDING GAS		BASE METAL	
Type	Size	Amps	Wire Feed Speed IPM (cm/min)	Volts	Type	Flow	Type	Thickness
E70S-3	0.035 in (0.9 mm)	90 to 120	180 to 300 (457 to 762)	15 to 19	CO_2 or 75%Ar/$CO_2$25%	30 to 50	Low-Carbon Steel	1/4 in to 1/2 in (6 mm to 13 mm)
E70S-3	0.045 in (1.2 mm)	130 to 200	125 to 200 (318 to 508)	17 to 20	CO_2 or 75%Ar/$CO_2$25%	30 to 50	Low-Carbon Steel	1/4 in to 1/2 in (6 mm to 13 mm)

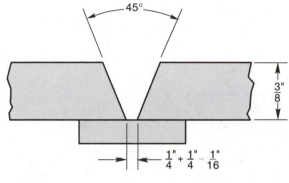

Figure 13-3

10. Electrical characteristics: The type of welding current (usually direct current electrode positive [DCEP]) is specified here.

    Item 10b lists the base metal polarity as being either negative or positive.

11. Operating range specifications: Table 13-3 is a detailed listing of the allowable ranges for electrode diameter gas flow rate, voltage, and amperage for each thickness of metal covered by this WPS. You can use this chart by first selecting the thickness of the material to be welded and then following horizontally across the page to the right, where, in each column, specific ranges are given for each welding parameter to be used in making the weld.

12. Preheat: Unless a higher preheat is required for the particular type of metal to be welded, a minimum temperature of 50° F (10° C) is the lowest acceptable temperature for welding.

13. Backing gas: On some critical welds, pipe welds, and welds made on reactive metals such as stainless steel, backing gas is used to surround the root face of the weld as it is being produced. This gas prevents contamination of the root face by the atmosphere. The specific type of gas, flow rate, purging procedures, and other detailed information are included in this section.

14. Safety: General and specific safety procedures and requirements for the shop and job are given here.

15. Welding techniques: Information about (15a) tack welds; such as size, location, and sequence of placement, are included as well as (15b) groove or fillet weld information, such as number of passes, weld bead sequences, bead sizes, reinforcement, and stopping and starting techniques may be included here.

16. Inner pass temperature: This is the maximum allowable temperature that the base metal may reach during welding. Long periods of elevated temperature can significantly affect the mechanical properties of most metals and also adversely affect the weld bead's shape. Thus it is important that this temperature be stated. Additionally, information regarding the method of cooling should be stated.

17. Cleaning: From time to time during the welding process, additional cleaning or recleaning of the weld may be required. In some cases (such as thick sections on mild steel) this cleaning may include some weld bead reshaping using grinders. Any such allowable operations must be outlined here.

18. Inspection: The method and extent of examination of the finished weld as well as tolerances of acceptability must be given.

19. Repair: If the weld fails during postwelding inspection, in many industrial applications specific repair procedures are allowed. Repair procedures may include methods of removing weld discontinuities and effects as well as allowable procedures for making such welding repairs. In many cases reference to a completely separate WPS for making such repairs will be noted here.

20. Sketches: A series of sketches or complete mechanical drawings may be included. Drawings that are included or that accompany a WPS would be for a specific part or weldment that would be fabricated using this specific procedure. The same WPS can be used on any number of products as long as they fall within the scope of the WPS (Figure 13-3).

JOINT DESIGN AND TOLERANCES:

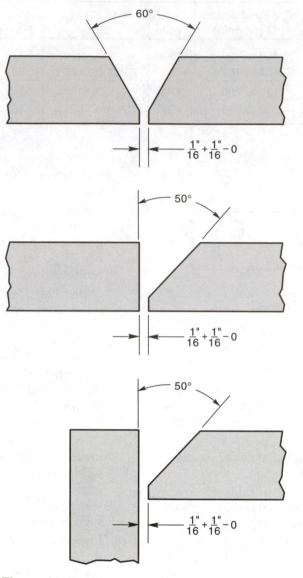

Figure 13-4

PREPARATION OF BASE METAL: The bevels are to be flame cut on the edges of the plate before the parts are assembled. The beveled surface must be smooth and free of notches. Any roughness or notches that are deeper than 1/64″ (0.4 mm) must be ground smooth.

All hydrocarbons and other contaminants, such as cutting fluids, grease, oil, and primers, must be cleaned off of all parts and filler metals before welding. This cleaning can be done with any suitable solvents or detergents. The groove face and inside and outside plate surfaces within 1″ (25 mm) of the joint must be mechanically cleaned of slag, rust, and mill scale. Cleaning must be done with a wire brush or grinder down to bright metal.

ELECTRICAL CHARACTERISTICS: The current shall be <u>Direct Current Electrode Positive (DCEP)</u>.
The base metal shall be on the <u>negative</u> side of the line.

ELECTRODE		WELDING POWER			SHIELDING GAS		BASE METAL	
Type	Size	Amps	Wire Feed Speed IPM (cm/min)	Volts	Type	Flow	Type	Thickness
E70S-3	0.035 in (0.9 mm)	90 to 120	180 to 300 (457 to 762)	15 to 19	CO_2 or 75%Ar/$CO_2$25%	30 to 50	Low-Carbon Steel	1/4 in to 1/2 in (6 mm to 13 mm)
E70S-3	0.045 in (1.2 mm)	130 to 200	125 to 200 (318 to 508)	17 to 20	CO_2 or 75%Ar/$CO_2$25%	30 to 50	Low-Carbon Steel	1/4 in to 1/2 in (6 mm to 13 mm)

PREHEAT: The parts must be heated to a temperature higher than 50° F (10° C) before any welding is started.

BACKING GAS: N/A

SAFETY: Proper protective clothing and equipment must be used. The area must be free of all hazards that may affect the welder or others in the area. The welding machine, welding leads, work clamp, electrode holder, and other equipment must be in safe working order.

WELDING TECHNIQUE: Using a 1/2″ (13 mm) or larger gas nozzle for all welding, first tack weld the plates together according to the drawing. There should be about a 1/16″ (1.6 mm) root gap between the plates with V-grooved or beveled edges. Use the E70S-3 arc welding electrodes to make a root pass to fuse the plates together. Clean any silicon slag from the root pass being, and remove any trapped silicon slag along the sides of the weld.

Using the E70S-3 arc-welding electrodes, make a series of stringer or weave filler welds no thicker than 1/4″ (6.4 mm) in the groove until the joint is filled. The 1/4″ (6.4 mm) fillet welds are to be made with one pass.

INTERPASS TEMPERATURE: The plate should not be heated to a temperature higher than 350° F (175° C) during the welding process. After each weld pass is completed, allow it to cool but never to a temperature below 50° F (10° C). The weldment must not be quenched in water.

CLEANING: Any slag must be removed between passes. The weld beads may be cleaned by a hand wire brush, hand chipping, punch and hammer, or a needle-scaler. All weld cleaning must be performed with the test plate in the welding position. A grinder may not be used to remove weld control problems such as undercut, overlap, or trapped slag.

INSPECTION: Visually inspect the weld for uniformity and discontinuities. There shall be no cracks, no incomplete fusion, and no overlap. Undercut shall not exceed the lesser of 10% of the base metal thickness or 1/32″ (0.8 mm). The frequency of porosity shall not exceed one in each 4″ (100 mm) of weld length, and the maximum diameter shall not exceed 3/32″ (2.4 mm).

REPAIR: No repairs of defects are allowed.

SKETCHES (Figure 13-5):

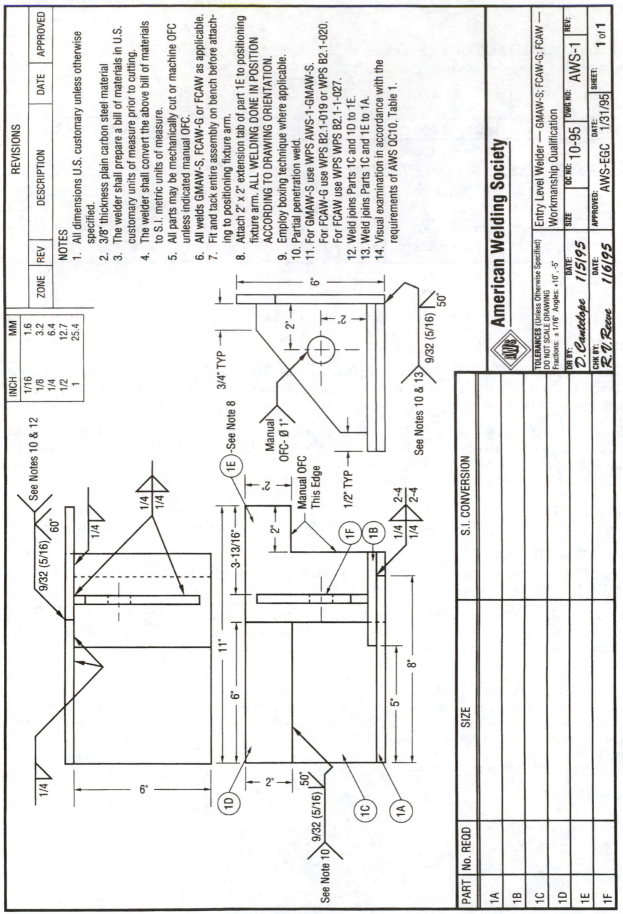

Figure 13-5 GMAW-S Workmanship Qualification Test. (Courtesy of the American Welding Society)

GAS METAL ARC WELDING-SHORT CIRCUIT METAL TRANSFER (GMAW-S)

WELDING PROCEDURE SPECIFICATION (WPS)
Welding Procedures Specifications No: TEW GMAW 2 _____ Date: _____

TITLE: Welding _GMAW-S_____ of _plate_____ to _plate_____.

SCOPE: This procedure is applicable for _V-groove, bevel and fillet welds_____ within the range of _24 ga. (0.52 mm)_ through _1/8 in. (3 mm)_. Welding may be performed in the following positions _ALL_____.

BASE METAL: The base metal shall conform to _Carbon Steel M-1, P-1 and S-1 Group 1 or 2_ Backing material specification _NONE_____.

FILLER METAL: The filler metal shall conform to AWS specification No. _E70S-1_____ from AWS specification _A5.18_____. This filler metal falls into F-number _F-6_____ and A-number _A-1_____.

SHIELDING GAS: The shielding gas, or gases, shall conform to the following compositions and purity: CO_2 at 30 to 50 CFH or 75% Ar/CO2 25% at 30 to 50 CFH.

JOINT DESIGN AND TOLERANCES:

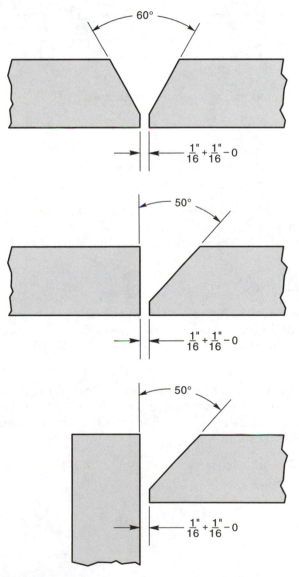

Figure 13-6

PREPARATION OF BASE METAL: The edge is to be sheared or plasma cut before the parts are assembled. The edge surface must be smooth and free of notches. Any roughness or notches that are deeper than 1/64″ (0.4 mm) must be ground smooth.

All hydrocarbons and other contaminations, such as cutting fluids, grease, oil, and primers, must be cleaned off of all parts and filler metals before welding. This cleaning can be done with any suitable solvents or detergents. The groove face and inside and outside plate surface within 1″ (25 mm) of the joint must be mechanically cleaned of slag, rust, and mill scale. Cleaning must be done with a wire brush or grinder down to bright metal.

ELECTRICAL CHARACTERISTICS: The current shall be Direct Current Electrode Positive (DCEP) . The base metal shall be on the negative side of the line.

ELECTRODE		WELDING POWER			SHIELDING GAS		BASE METAL	
Type	Size	Amps	Wire Feed Speed IPM (cm/min)	Volts	Type	Flow	Type	Thickness
E70S-1	0.030 in (0.8 mm)	60 to 140	150 to 350 (381 to 889)	14 to 16	CO_2 or Ar + CO_2	30 to 50	Low-Carbon Steel	16 gal to 1/8 in (1.5 mm to 3 mm)
E70S-1	0.035 in (0.9mm)	90 to 160	180 to 300 (457 to 762)	15 to 19	CO_2 or Ar + CO_2	30 to 50	Low-Carbon Steel	16 gal to 1/8 in (1.5 mm to 3 mm)

PREHEAT: The parts must be heated to a temperature higher than 50° F (10° C) before any welding is started.

BACKING GAS: N/A

SAFETY: Proper protective clothing and equipment must be used. The area must be free of all hazards that may affect the welder or others in the area. The welding machine, welding leads, work clamp, electrode holder, and other equipment must be in safe working order.

WELDING TECHNIQUE: Using a 1/2″ (13 mm) or larger gas nozzle for all welding, first tack weld the plates together according to the drawing. There should be about a 0″ (0-mm) to 1/16″ (1.6 mm) root gap between the plates edges. Use the E70S-1 arc welding electrodes to make a single pass to fuse the plates together. Clean any silicon slag from the pass.

INTERPASS TEMPERATURE: N/A

CLEANING: Any slag must be cleaned off passes. The weld beads may be cleaned by a hand wire brush, hand chipping, punch and hammer, or a needle-scaler. All weld cleaning must be performed with the test plate in the welding position. A grinder may not be used to remove weld control problems such as undercut, overlap, and trapped slag.

INSPECTION: Visually inspect the weld for uniformity and discontinuities. There shall be no cracks, no incomplete fusion and no overlap. Undercut shall not exceed the lesser of 10% of the base metal thickness or 1/32 in. (0.8 mm). The frequency of porosity shall not exceed one in each 4 in. (100 mm) of weld length and the maximum diameter shall not exceed 3/32 in. (2.4 mm).

REPAIR: No repairs of defects are allowed.

SKETCHES (Figure 13-7):

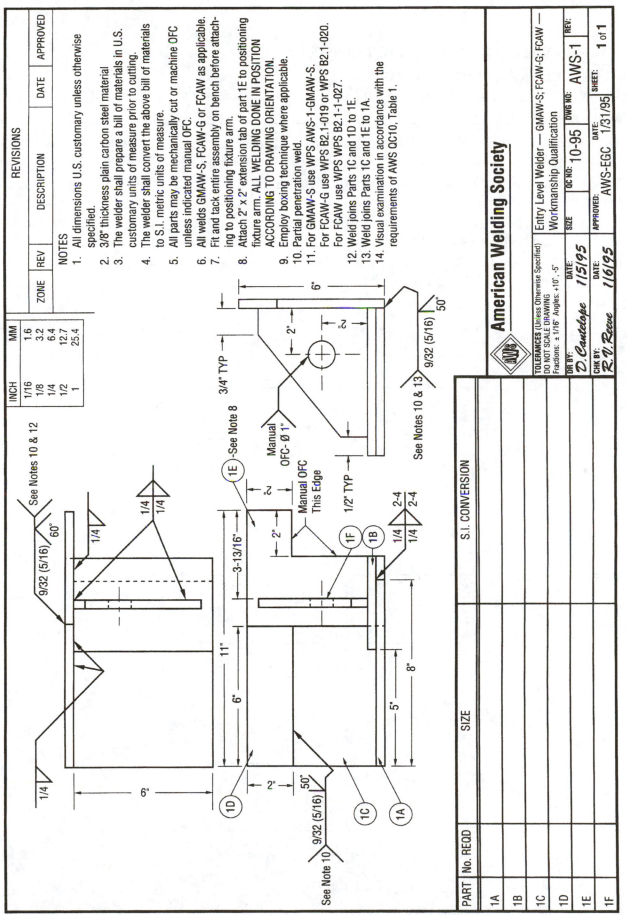

REVISIONS		
ZONE	REV	DESCRIPTION	DATE	APPROVED

NOTES

1. All dimensions U.S. customary unless otherwise specified.
2. 3/8" thickness plain carbon steel material
3. The welder shall prepare a bill of materials in U.S. customary units of measure prior to cutting.
4. The welder shall convert the above bill of materials to S.I. metric units of measure.
5. All parts may be mechanically cut or machine OFC unless indicated manual OFC.
6. All welds GMAW-S, FCAW-G or FCAW as applicable.
7. Fit and tack entire assembly on bench before attaching to positioning fixture arm.
8. Attach 2" x 2" extension tab of part 1E to positioning fixture arm. ALL WELDING DONE IN POSITION ACCORDING TO DRAWING ORIENTATION.
9. Employ boxing technique where applicable.
10. Partial penetration weld.
11. For GMAW-S use WPS AWS-1-GMAW-S. For FCAW-G use WPS B2.1-019 or WPS B2.1-020. For FCAW use WPS WPS B2.1-1-027.
12. Weld joins Parts 1C and 1D to 1E.
13. Weld joins Parts 1C and 1E to 1A.
14. Visual examination in accordance with the requirements of AWS QC10, Table 1.

INCH	MM
1/16	1.6
1/8	3.2
1/4	6.4
1/2	12.7
1	25.4

American Welding Society

TOLERANCES (Unless Otherwise Specified) DO NOT SCALE DRAWING Fractions: ± 1/16" Angles: +10°, -5°	Entry Level Welder — GMAW-S; FCAW-G; FCAW — Workmanship Qualification	
DR BY: D. Cantelope 1/5/95	SIZE	QC NO: 10-95	DWG NO: AWS-1	REV:
CHK BY: R. V. Reeve 1/6/95	APPROVED: AWS-EGC	DATE: 1/31/95	SHEET: 1 of 1

PART	No. REQD	SIZE	S.I. CONVERSION
1A			
1B			
1C			
1D			
1E			
1F			

Figure 13-7 GMAW-S Workmanship Qualification Test. (Courtesy of the American Welding Society)

141

GAS METAL ARC WELDING-SPRAY METAL TRANSFER

WELDING PROCEDURE SPECIFICATION (WPS)

Welding Procedures Specifications No: <u>TEW GMAW 3</u>   Date: _____

TITLE: Welding <u>GMAW-S</u>   of <u>plate</u>   to <u>plate</u>   .

SCOPE: This procedure is applicable for <u>V-groove, bevel and fillet welds</u>   within the range of <u>1/8 in. (3.2 mm)</u> through <u>1 1/2 in. (38 mm)</u>. Welding may be performed in the following positions <u>ALL</u>   .

BASE METAL: The base metal shall conform to <u>Carbon Steel M-1, P-1 and S-1 Group 1 or 2</u> Backing material specification <u>NONE</u>   .

FILLER METAL: The filler metal shall conform to AWS specification No. <u>E70S-3</u>   from AWS specification <u>A5.18</u>   . This filler metal falls into F-number <u>F-6</u>   and A-number <u>A-1</u>   .

SHIELDING GAS: The shielding gas, or gases, shall conform to the following compositions and purity: <u>Argon with 2 to 5% Oxygen 25% at 30 to 50 CFH</u>   .

JOINT DESIGN AND TOLERANCES:

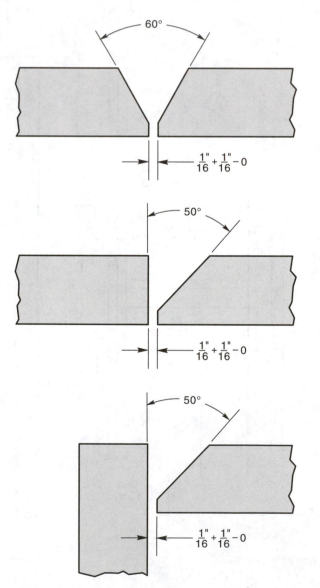

Figure 13-8

PREPARATION OF BASE METAL: The bevels are to be flame cut on the edges of the plate before the parts are assembled. The beveled surface must be smooth and free of notches. Any roughness or notches that are deeper than 1/64″ (0.4 mm) must be ground smooth.

All hydrocarbons and other contaminations, such as cutting fluids, grease, oil and primers, must be cleaned off of all parts and filler metals before welding. This cleaning can be done with any suitable solvents or detergents. The groove face and inside and outside plate surfaces within 1″ (25 mm) of the joint must be mechanically cleaned of slag, rust, and mill scale. Cleaning must be done with a wire brush or grinder down to bright metal.

ELECTRICAL CHARACTERISTICS: The current shall be <u>Direct Current Electrode Positive (DCEP)</u>. The base metal shall be on the <u>negative</u> side of the line.

ELECTRODE		WELDING POWER			SHIELDING GAS		BASE METAL	
Type	Size	Amps	Wire Feed Speed IPM (cm/min)	Volts	Type	Flow	Type	Thickness
E70S-3	0.035 in (0.9 mm)	180 to 230	400 to 550 (1016 to 1397)	25 to 27	Ar + 2% to 5% C$_2$	30 to 50	Low-Carbon Steel	1/4 in to 1/2 in (6 mm to 13 mm)
E70S-3	0.045 in (1.2 mm)	260 to 340	300 to 500 (762 to 1270)	25 to 30	Ar + 2% to 5% C$_2$	30 to 50	Low-Carbon Steel	1/4 in to 1/2 in (6 mm to 13 mm)

PREHEAT: The parts must be heated to a temperature higher than 50° F (10° C) before any welding is started.

BACKING GAS: N/A

SAFETY: Proper protective clothing and equipment must be used. The area must be free of all hazards that may affect the welder or others in the area. The welding machine, welding leads, work clamp, electrode holder, and other equipment must be in safe working order.

WELDING TECHNIQUE: Using a 1/2″ (13 mm) or larger gas nozzle for all welding, first tack weld the plates together according to the drawing. There should be about a 1/16″ (1.6 mm) root gap between the plates with V-grooved or beveled edges. Use the E70S-3 arc welding electrodes to make a root pass to fuse the plates together. Clean any silicon slag from the root pass being sure to remove any trapped silicon slag along the sides of the weld.

Using the E7OS-3 arc welding electrodes, make a series of stringer or weave filler welds, no thicker than 1/4″ (6.4 mm), in the groove until the joint is filled. The 1/4″ (6.4 mm) fillet welds are to be made with one pass.

INTERPASS TEMPERATURE: The plate should not be heated to a temperature higher than 350° F (175° C) during the welding process. After each weld pass is completed, allow it to cool but never to a temperature below 50° F (10° C). The weldment must not be quenched in water.

CLEANING: Any slag must be cleaned off between passes. The weld beads may be cleaned by a hand wire brush, a hand chipping, punch and hammer, or a needle-scaler. All weld cleaning must be performed with the test plate in the welding position. A grinder may not be used to remove weld control problems such as undercut, overlap, and trapped slag.

INSPECTION: Visually inspect the weld for uniformity and discontinuities. There shall be no cracks, no incomplete fusion, and no overlap. Undercut shall not exceed the lesser of 10% of the base metal thickness or 1/32″ (0.8 mm). The frequency of porosity shall not exceed one in each 4″ (100 mm) of weld length, and the maximum diameter shall not exceed 3/32″ (2.4 mm).

REPAIR: No repairs of defects are allowed.

SKETCHES (Figure 13-9):

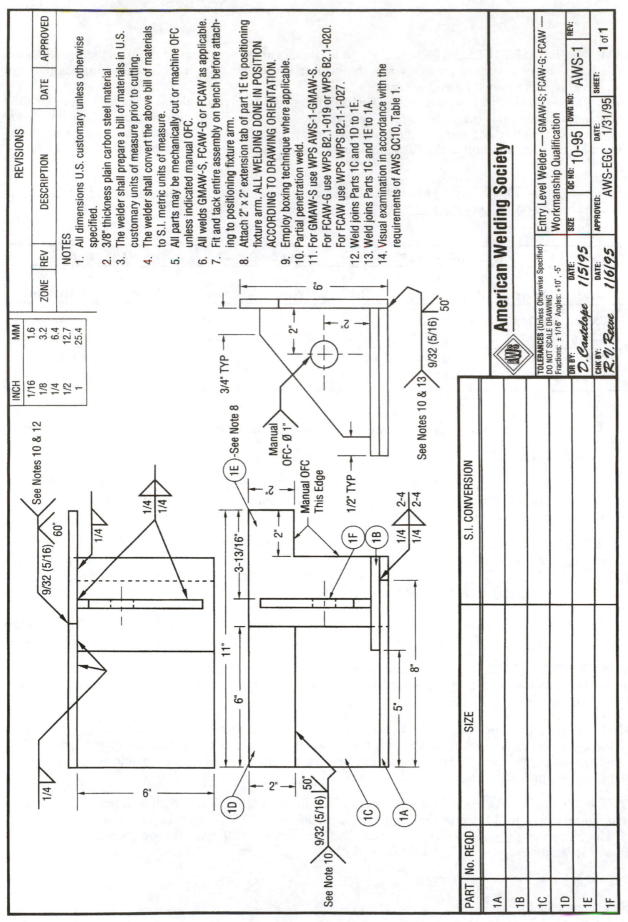

Figure 13-9 GMAW-S Workmanship Qualification Test. (Courtesy of the American Welding Society)

SUMMARY

Welding standards that have been established for particular welds as the result of testing significantly reduce problems associated with the improper setup of the welding equipment. There is still a need to have skilled welders to do the actual welding.

REVIEW QUESTIONS

1. What is a WPS?
2. What information should a welding procedure specification contain?
3. What information does the title include?
4. What information is included for preparation of base metal?
5. What type of welding current is usually used?
6. When is backing gas used?
7. What is the *inner pass temperature*?

CHAPTER 14

WELD TESTING AND INSPECTION

A welder must know that a weld will meet the requirements of a company and/or codes and standards. Weld testing and inspection ensure that the welds produced meet the quality, reliability, and strength requirements of the weldments. The extent to which welders and products are subjected to testing and inspection depends on the requirements of the industry. Many welding products are used in noncritical applications, such as those in which the failure of the weld would result in only a minor inconvenience. Critical applications, on the other hand, might occur in industry, where the failure of a weld could result in severe loss of property and life. Noncritical welds can be found in applications such as ornamental furniture, sculptures, jigs and fixtures, and other household or industrial products. Critical welds are welds that would be produced in, for example, the nuclear power industry, oil refinery and chemical industries, and the aircraft and spacecraft industries.

The level and intensity of testing and inspection are greatly influenced by the weld's intended purpose. The more intense the evaluation process, the more expensive the weld and weldment are to produce. It is therefore important that welds and weldments be accepted only to standards that are necessary to ensure that the finished product is fit for service.

QUALITY CONTROL (QC)

The first step in establishing a quality-control program is to determine the appropriate codes and standards for the weldments being produced.

The selected code or standard will then determine the type of testing. Two major classifications of testing products are *destructive* and *nondestructive testing*. Nondestructive testing can be performed on a product without affecting the product in any way. Destructive testing renders the weldment unusable. Nondestructive testing is often used to both qualify the welding procedure and ensure the quality of the weldment. During the establishment and verification of the welding procedure specifications, destructive testing may be used in conjunction with nondestructive examination. Destructive testing is used to establish the acceptable criteria for the nondestructive testing that will be applied to the products.

Destructive or mechanical testing is used as a baseline evaluation to establish welding procedure specifications. It may also be used randomly on completed products to ensure that the weld quality is being maintained.

DISCONTINUITIES AND DEFECTS

A discontinuity is an interruption of the typical structure of a weldment. It may be a lack of uniformity in the mechanical, metallurgical, or physical characteristics of the material or weldment. A discontinuity is not necessarily a defect.

A defect, according to AWS, is "a discontinuity or discontinuities, which by nature or accumulated effect (for example, total porosity or slag inclusion length) that renders a part or product unable to meet minimum applicable acceptance standards or specifications."

In other words many acceptable products may have welds that contain discontinuities. But no products may have welds that contain defects. The only difference between a *discontinuity* and a *defect* is when the discontinuity becomes unacceptably large or when there are so many small discontinuities that the weld is not acceptable under the standards of the code for that product. Some codes are stricter than others, so the same weld might be acceptable under one code but not under another.

Ideally, a weld should not have any discontinuities, but that is practically impossible. The difference between what is acceptable, fit for service, and perfection is known as *tolerance*. In many industries, the tolerances for welds have been established and are available as codes or standards. Table 14-1 lists a few of the agencies that issue codes or standards. Each code or standard gives the tolerance that determines when a discontinuity becomes a defect.

When evaluating a weld, it is important to note the type, the size, and the location of the discontinuity. Any one of these factors or all three can be the deciding factor that, based on the applicable code or standard, might change a discontinuity into a defect.

TABLE 14-1

WELDING CODES AND SPECIFICATIONS

A *welding code* is a detailed listing of the rules or principles that are to be applied to a specific classification or type of product.

A *welding specification* is a detailed statement of the legal requirements for a specific classification or type of product. Products manufactured to code or specification requirements commonly must be inspected and tested to ensure compliance.

A number of agencies and organizations publish welding codes and specifications. The application of the particular code or specification to a weldment can be the result of one or more of the following requirements:

- Local, state, or federal government regulations
- Bonding or insuring company
- End user (customer) requirements
- Standard industrial practices

The three most popular codes are:

#1104, American Petroleum Institute—used for pipelines

Section IX, American Society of Mechanical Engineers—used for pressure vessels

D1.1, American Welding Society—used for bridges and buildings

The following organizations publish welding codes and/or specifications:

AASHT
    American Association of State Highway and Transportation Officials
    444 North Capitol Street, NW
    Washington, DC 20001

AISC
    American Institute of Steel Construction
    1 East Wacker Drive
    Chicago, IL 60601

ANSI
    American National Standards Institute
    11 W. 42nd Street
    New York, NY 10036

API
    American Petroleum Institute
    2101 L Street, NW
    Washington, DC 20005

AREA
    American Railway Engineering Association
    50 F. Street, NW
    Washington, DC 20001

ASME
    American Society of Mechanical Engineers
    345 East 47th Street
    New York, NY 10017

AWWA
    American Water Works Association
    6666 West Quincy Avenue
    Denver, CO 80235

AWS
    American Welding Society
    550 NW LeJeune Road
    Miami, FL 33126

AAR
    Association of American Railroads
    50 F. Street, NW
    Washington, DC 20001

MIL
    Department of Defense
    Washington, DC 20301

SAE
    Society of Automotive Engineers
    400 Commonwealth Drive
    Warrendale, PA 15096

The twelve most common discontinuities are as follows:

- Porosity
- Inclusions
- Inadequate joint penetration
- Incomplete fusion
- Arc strikes
- Overlap
- Undercut
- Cracks
- Underfill
- Lamination
- Delamination
- Lamellar tears

Porosity

Porosity results from gas that was dissolved in the molten weld pool, forming bubbles that are trapped as the metal cools and solidifies. The bubbles that make up porosity form within the weld metal, and for that reason they cannot be seen as they form. These gas pockets form in the same way that bubbles form in a carbonated drink as it warms up or as air dissolved in water forms bubbles in the center of a cube of ice.

Porosity bubbles are either spherical (ball-shaped) or cylindrical (tube- or tunnel-shaped). Cylindrical porosity is called a *wormhole*. The rounded edges tend to reduce the stresses around them. Therefore, unless porosity is extensive, there is little or no loss in strength.

Porosity is most often caused by improper welding techniques, contamination, or an improper chemical balance between the filler and base metals.

Improper welding techniques may result in shielding gas not properly protecting the molten weld pool. For example, drafts may blow the shielding gas coverage away from the molten weld pool, or inadequate prepurging may leave some atmosphere in the welding zone during start-up.

Often with gas tungsten arc welding, welders try to resolve a porosity problem by remelting the weld. Welders assume that this is possible because additional filler metal need not be added, thus preventing what would appear to be an oversized weld from being produced. This practice is unacceptable because there is no way of removing the dissolved gases, and porosity thus occurs. Remelting simply results in these gases forming porosity below the surface or being dissolved in the weld metal. Such dis-solved gases will result in significant reduction in the mechanical properties, strength, ductility, and hardness of the weld.

- Uniformly scattered porosity is most frequently caused by poor welding techniques or faulty materials (Figure 14-1).
- Clustered porosity is most often caused by improper starting and stopping techniques (Figure 14-1).
- Linear porosity is most frequently caused by contamination within the joint, root, or inter-bead boundaries (Figure 14-1).
- Piping porosity, or a wormhole, is most often caused by contamination at the root (Figure 14-1). This porosity is unique because its formation depends on the gas escaping from the weld pool at the same rate as the pool is solidifying.

Inclusions

There are two major classifications of inclusions; metallic and nonmetallic. Nonmetallic inclusions are produced when materials such as slag become trapped in the weld metal, often between weld beads. Since GMA welding does not use slag-producing fluxes, nonmetallic inclusions are rare for GMA welding.

Inadequate Joint Penetration

Inadequate joint penetration occurs when the depth that the weld penetrates the joint is less than that needed to fuse through the plate or into the preceding weld (Figure 14-2). A defect usually results that could reduce the required cross-sectional area of the joint or become a source of stress concentration that leads to fatigue failure. The importance of such defects depends on the notch sensitivity of the metal and the factor of safety to which the weldment has been designed. Generally, if proper welding procedures were developed and followed, such defects will not occur.

The major causes of inadequate joint penetration are the following:

- Improper welding technique–The most common cause is a misdirected arc. Also, the welding technique may require that both starting and run-out tabs be used so that the molten weld pool is well established before it reaches the joint. Sometimes a failure to back gouge the root sufficiently provides a deeper root face than allowed for.

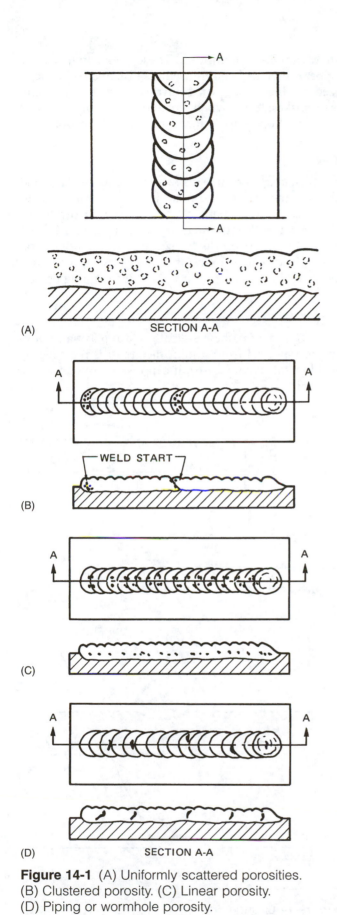

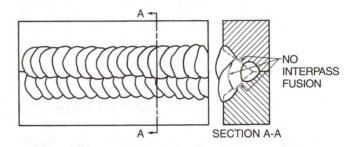

Figure 14-2 Inadequate joint penetration.

Figure 14-3 Incomplete fusion.

Figure 14-1 (A) Uniformly scattered porosities. (B) Clustered porosity. (C) Linear porosity. (D) Piping or wormhole porosity.

- Not enough welding current–Metals that are thick or have a high thermal conductivity are often preheated. Thus the weld heat is not drawn away so quickly by the metal that it cannot penetrate the joint.
- Improper joint fit-up–This problem results when the weld joints are not accurately prepared or fitted. Too small a root gap or too large a root face will keep the weld from penetrating adequately.
- Improper joint design–When joints are accessible from both sides, back gouging is often used to ensure 100% root fusion.

Incomplete Fusion

Incomplete fusion is the lack of coalescence between the molten filler metal and previously deposited filler metal and/or the base metal (Figure 14-3). The lack of fusion between the filler metal and previously deposited weld metal is called *interpass cold lap*. The lack of fusion between the weld metal and the joint face is called *lack of sidewall fusion*. Both of these problems usually travel along all or most of the weld's length. This discontinuity is not as detrimental to the weld's strength in service if it is near the center of the weld and is not open to the surface.

Some major causes of lack of fusion are the following:

- Inadequate agitation–Lack of weld agitation to break up oxide layers. The base metal or weld filler metal may melt, but a thin layer of oxide may prevent coalescence.

- Improper welding techniques–Poor manipulation, such as moving too fast.

- Improper edge preparation–All notches and gouges in the edge of the weld joint must be removed. For example, if a flame-cut plate has notches along the cut, they could cause a lack of fusion in each notch (Figure 14-4).

- Improper joint design–Incomplete fusion may also result from not enough heat to melt the base metal or too little space allowed by the joint designer for correct molten weld pool manipulation.

- Improper joint cleaning–Failure to clean oxides for the joint surfaces resulting from the use of an oxyfuel torch to cut the plate or failure to remove slag from a previous weld. Incomplete fusion can be found in welds produced by all major welding processes.

Arc Strikes

Arc strikes are small, localized points where surface melting occurred away from the joint. These spots may be caused by accidentally striking the arc in the wrong place and/or by faulty ground connections. Even though arc strikes can be ground smooth, they cannot be removed. These spots will always appear if an acid etch is used. They can also be localized hardness zones or a starting point for cracking. Arc strikes, even when ground flush for a guided bend, will open up to form small cracks or holes.

Overlap

Overlap occurs in fusion welds when weld deposits are larger than the joint is conditioned to accept. The weld metal then flows over a surface of the base metal without fusing to it (Figure 14-5). It generally occurs on the horizontal leg of a horizontal fillet weld under extreme conditions. It can also occur on both sides of flat-positioned capping passes.

Undercut

Undercut is the result of the arc plasma removing more metal from a joint face than is replaced by weld metal (Figure 14-6). It can result from excessive current.

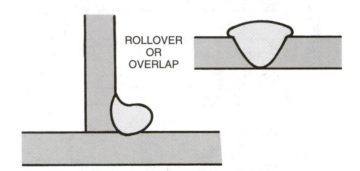

Figure 14-5 Rollover or overlap.

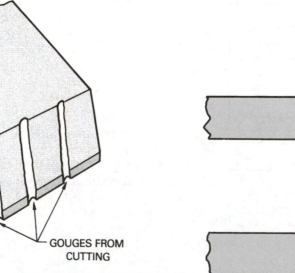

Figure 14-4 Remove gouges along the surface of the joint before welding.

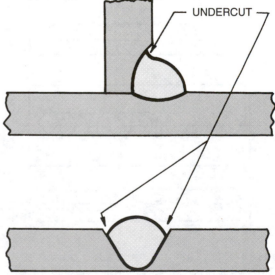

Figure 14-6 Undercut.

Crater Cracks

Crater cracks are the tiny cracks that develop in the weld craters as the weld pool shrinks and solidifies (Figure 14-7). Low-melting materials are rejected toward the crater center while freezing. Since these materials are the last to freeze, they are pulled apart or separate as a result of the weld metal shrinking as it cools, leaving crater cracks. The high-shrinkage stresses aggravate crack formation. Crater cracks can be minimized, if not prevented, by not interrupting the arc quickly at the end of a weld. This allows the arc to lengthen, the current to drop gradually, and the crater to fill and cool more slowly. GMAW equipment has a crater-filling control that automatically and gradually reduces the wire-feed speed at the end of a weld.

Underfill

Underfill on a groove weld occurs when the weld metal deposited is inadequate to bring the weld's face or root surfaces to a level, equal to that of the original plane. For a fillet weld underfill occurs when the weld deposit has an insufficient effective throat (Figure 14-8).

Plate-Generated Problems

Not all welding problems are caused by weld metal, the process, or the welder's lack of skill in depositing that metal. The material being fabricated can be at fault, too. Some problems result from internal plate defects that the welder cannot control. Others are the result of improper welding procedures that produce undesirable hard metallurgical structures in the heat-affected zone, as discussed in other chapters. The internal defects are the result of poor steel-making practices. Steel producers try to keep their steels as sound as possible, but the mistakes that occur are blamed too frequently on the welding operation.

Lamination

Lamination differs from lamellar tearing because it is more extensive and involves thicker layers of nonmetallic contaminants. Located toward the center of the plate (Figure 14-9), lamination is caused by insufficient *cropping* (removal of defects) of the pipe in ingots. The slag and oxidized steel in the pipe are rolled out with the steel, producing the lamination.

Delamination

When lamination intersects a joint being welded, some lamination may open up and become delaminate. Contamination of the weld metal may occur if the lamination contained large amounts of slag, mill scale, dirt, or other undesirable materials. Such contamination can cause wormhole porosity or lack-of-fusion defects.

The problems associated with delaminations are not easily corrected. An effective solution for thick plate is to weld over the lamination to seal it. A better solution is to replace the steel.

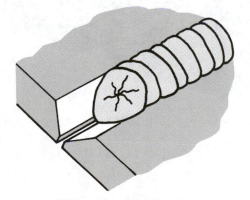

Figure 14-7 Crater or star cracks.

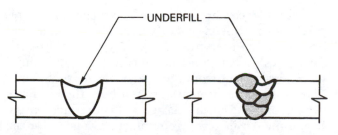

Figure 14-8

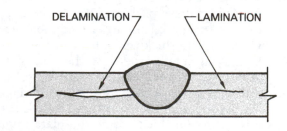

Figure 14-9

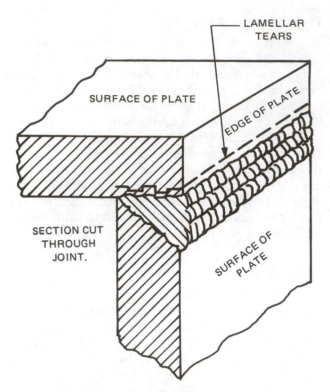

Figure 14-10 Example of lamellar tearing.

Figure 14-11 Using multiple welds to reduce weld stresses.

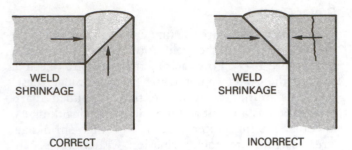

Figure 14-12 Correct joint design to reduce lamellar tears.

Figure 14-13 Typical tensile tester for measuring the strength of welds (60,000 lb universal testing machines) (Courtesy of Tinius Olsen Testing Machine Co., Inc.)

Lamellar Tears

These tears appear as cracks parallel to and under the steel surface. In general, they are not in the heat-affected zone, and they have a steplike configuration. They result from the thin layers of nonmetallic inclusions that lie beneath the plate surface and have very poor ductility. Although barely noticeable, these inclusions separate when severely stressed, producing laminated cracks. These cracks are evident if the plate edges are exposed (Figure 14-10).

A solution to the problem is to redesign the joints in order to impose the lowest possible strain throughout the plate thickness. This can be accomplished by making smaller welds, so that each subsequent weld pass heat-treats the previous pass to reduce the total stress in the finished weld (Figure 14-11). The joint design can be changed to reduce the stress on the through thickness of the plate (Figure 14-12).

DESTRUCTIVE TESTING (DT)

Tensile Testing

Tensile tests are performed with specimens prepared as round bars or flat strips. The simple round bars are often used for testing only the weld metal, sometimes called "all-weld metal testing." Round specimens are cut from the center of the weld metal. The flat bars are often used to test both the weld and the surrounding metal. Bar size also depends on the size of the tensile testing equipment available for the testing (Figure 14-13).

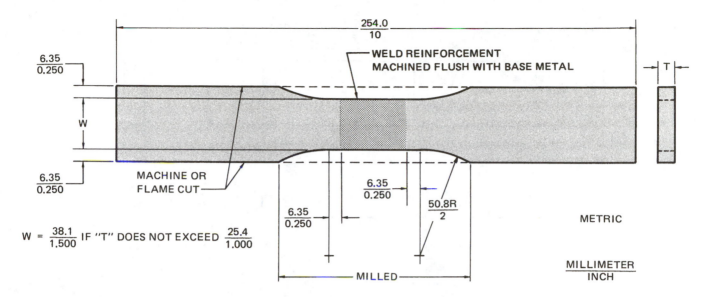

Figure 14-14 Tensile specimen for flat-plate weld. (Courtesy of Hobart Brothers Company, Troy, Ohio)

Two flat specimens are commonly used for testing thinner sections of metal. When you test welds, the specimen should include the heat-affected zone and the base plate. If the weld metal is stronger than the plate, failure occurs in the plate; if the weld is weaker, failure occurs in the weld. This test, then, is open to interpretation.

After the weld section is machined to the specified dimensions, it is placed in the tensile testing machine and pulled apart. A specimen used to determine the strength of a welded butt joint for plate is shown in Figure 14-14.

Fatigue Testing

Fatigue testing determines how well a weld can resist repeated fluctuating stresses or cyclic loading. The maximum value of the stresses is less than the tensile strength of the material.

In a fatigue test, a part is subjected to repeated changes in applied stress. This test may be performed in one of several ways, depending upon the type of service the tested part must withstand. The results obtained are usually reported as the number of stress cycles that the part will resist without failure and the total stress used.

Nick-Break Test

A specimen for a *nick-break* test is prepared as shown in Figure 14-15(A). The specimen is supported as shown in Figure 14-15(B). A force is then applied, and the specimen is ruptured by one or more blows of a hammer. The force may be applied slowly or suddenly. Theoretically, the rate of application can affect how the specimen breaks, especially at a critical temperature. Generally, however, there is no difference in the appearance of the fractured surface due to the method of applying force. The surfaces of the fracture should be checked for soundness of the weld.

Guided Bend Test

To test welded, grooved butt joints on metal that is 3/8″ (10 mm) thick or less, two specimens are prepared and tested—one face bend and one root bend (Figure 14-16[A] and [B]). If the welds pass this test, the welder is qualified to make groove welds on plate having a thickness range of from 3/8″ to 3/4″ (10 mm to 19 mm). These welds need to be machined as shown in Figure 14-17(A) and (B). If these specimens pass, the welder will also be qualified to make fillet welds on materials of any (unlimited) thicknesses. For welded, grooved butt joints on metal 1″ (25 mm) thick, two side-bend specimens are prepared and tested (Figure 14-17[C]). If the welds pass this test, the welder is qualified to weld on metals of unlimited thickness.

When the specimens are prepared, caution must be taken to ensure that all grinding marks run longitudinally to the specimen so that they do not cause stress cracking. In addition, the edges must be rounded to reduce cracking that tends to radiate from sharp edges.

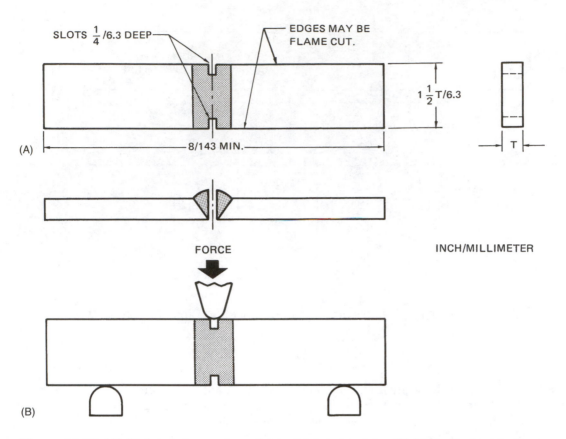

Figure 14-15 (A) Nick-break specimen for butt joints in plate. (B) Method of rupturing nick-break specimen. (Courtesy of Hobart Brothers Company, Troy, Ohio)

Testing by Etching

Specimens are etched for two purposes: (1) to determine the soundness of a weld, or (2) to determine the location of a weld.

A test specimen is produced by cutting a portion from the welded joint so that a complete cross-section is obtained. The face of the cut is then filed and polished with fine abrasive cloth. The specimen can then be placed in the etching solution. The etching solution or reagent makes the boundary between the weld metal and base metal visible (if the boundary is not already distinctly visible).

The most commonly used etching solutions are hydrochloric acid, ammonium persulphate, and nitric acid.

Hydrochloric Acid. Equal parts by volume of concentrated hydrochloric (muriatic) acid and water are mixed. The welds are immersed in the reagent at or near the boiling temperature. The acid usually enlarges gas pockets and dissolves slag inclusions, enlarging the resulting cavities.

CAUTION

When mixing the muriatic acid into the water, be sure to wear safety glasses and gloves to prevent injuries.

Ammonium Persulphate. A solution is prepared consisting of one part of ammonium persulphate (solid) to nine parts of water by weight. The surface of the weld is rubbed with cotton saturated with this reagent at room temperature.

Nitric Acid. A great deal of care should be exercised when using nitric acid because severe burns can result if it is used carelessly. One part of concentrated nitric acid is mixed with nine parts of water by volume.

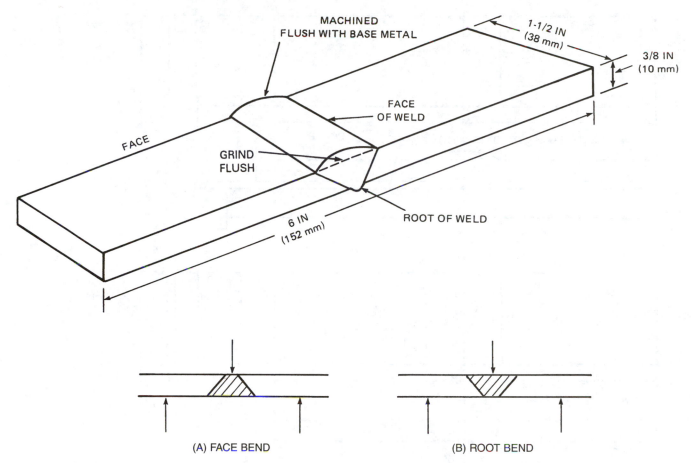

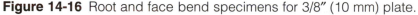

Figure 14-16 Root and face bend specimens for 3/8″ (10 mm) plate.

CAUTION

Always pour the acid slowly into the water, while continuously stirring the water, when diluting. Careless handling of this material or pouring *water into the acid* can result in burns, excessive fuming, or explosion.

The reagent is applied to the surface of the weld with a glass stirring rod at room temperature. Nitric acid has the capacity to etch rapidly and should be used on polished surfaces only.

After etching, the weld is rinsed in clear, hot water. Excess water is removed, and the etched surface is then immersed in ethyl alcohol and dried.

Impact Testing

A number of tests can be used to determine the impact capability of a weld. One common test is the *Izod test* (Figure 14-18[A]), in which a notched specimen is struck by an anvil mounted on a pendulum.

The energy in footpounds required to break the specimen indicates the impact resistance of the metal. This test compares the toughness of the weld metal with the base metal.

Another type of impact test is the *Charpy test,* which is similar to the Izod test. The difference is in the manner in which specimens are held. A typical impact tester is shown in Figure 14-18(B).

NONDESTRUCTIVE TESTING (NDT)

Nondestructive testing of welds is a method used to test materials for surface defects such as cracks, arc strikes, undercuts, and lack of penetration. Internal or subsurface defects can include tungsten inclusions, porosity, and unfused metal in the interior of the weld.

Visual Inspection (VT)

Visual inspection is the most frequently used nondestructive testing method. The majority of welds

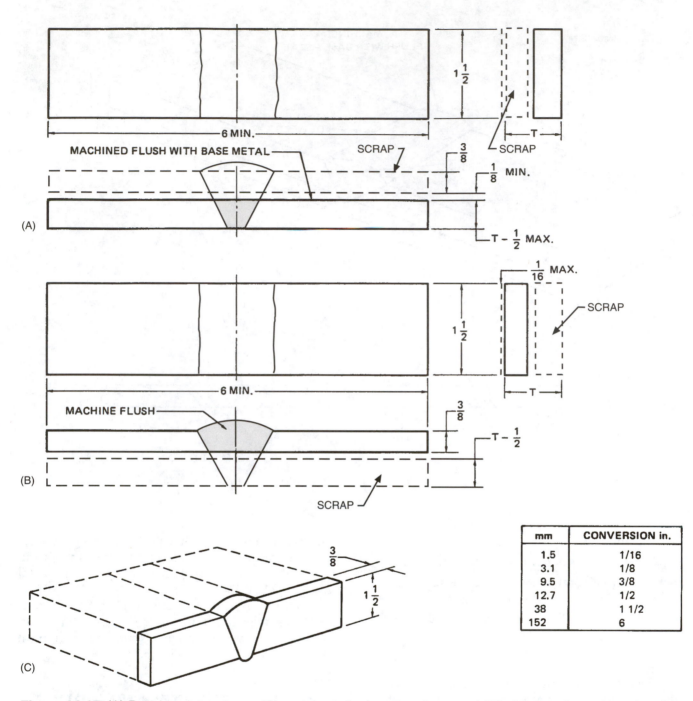

mm	CONVERSION in.
1.5	1/16
3.1	1/8
9.5	3/8
12.7	1/2
38	1 1/2
152	6

Figure 14-17 (A) Root bend specimen, (B) specimen for face bend test and (C) side-bend specimen for plates thicker than 3/8″ (10 mm).

receive only visual inspection. In this method, if the weld looks good, it passes; if it looks bad, it is rejected. Unfortunately, this procedure is often overlooked when more sophisticated nondestructive testing methods are used.

Visual inspection can easily be used to check for fit-up, interpass acceptance, welder technique, and other variables that affect a weld's quality. Minor problems can be identified and corrected before a weld is completed, eliminating costly repairs or rejection.

Visual inspection should be used before any other nondestructive or mechanical tests to eliminate (reject) obvious problem welds. Eliminating welds

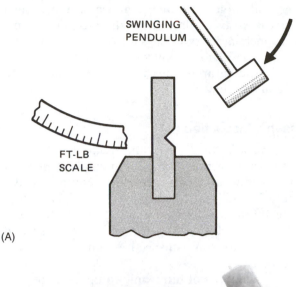

Figure 14-18 Impact testing: (A) specimen mounted for Izod impact toughness and (B) typical impact tester for measuring the toughness of metals. (Photo courtesy of Tinius Olsen Testing Machine Co., Inc.)

Figure 14-19 (Adapted from Magnaflux Corporation)

that have excessive surface discontinuities that will not pass the code or standards being used saves preparation time.

Penetrant Inspection (PT)

Penetrant inspection is used to locate minute surface cracks and porosity. Two types of penetrants are now in use: the color-contrast type and the fluorescent version. Color-contrast (often red) penetrants contain a colored dye that shows up under ordinary white light. Fluorescent penetrants contain a more effective fluorescent dye that shows up under black light.

Magnetic Particle Inspection (MT)

Magnetic particle inspection uses finely divided ferromagnetic particles (powder) to indicate defects on magnetic materials.

A magnetic field is induced in a part by passing an electric current through or around it. The magnetic field is always at right angles to the direction of current flow. Ferromagnetic powder registers an abrupt change in the resistance in the path of the magnetic field, such as would be caused by a crack lying at an angle to the direction of the magnetic poles at the crack. Finely divided ferromagnetic particles applied to the area will be attracted to and outline the crack.

In Figure 14-19, the flow or discontinuity interrupting the magnetic field in a test part can be either longitudinal or circumferential. A different type of magnetization is used to detect defects that run down the axis, as opposed to those occurring around the girth of a part.

Radiographic Inspection (RT)

Radiographic inspection is a method for detecting flaws inside weldments. Instead of using visible light

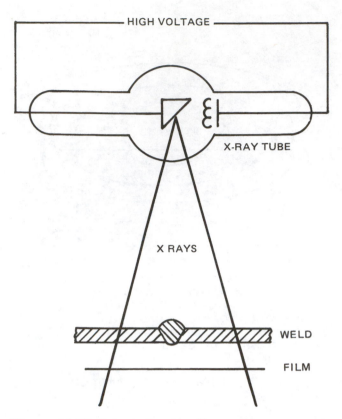

HIGH VOLTAGE

X-RAY TUBE

X RAYS

WELD

FILM

Figure 14-20 Schematic of an X-ray system.

rays, the operator uses invisible, short-wavelength rays developed by X-ray machines, radioactive isotopes (gamma rays), and variations of these methods. These rays are capable of penetrating solid materials and reveal most flaws in a weldment on an X-ray film or fluorescent screen. Flaws are revealed on films as dark or light areas against a contrasting background after exposure and processing (Figure 14-20).

Defect images in radiographs measure differences in how the X-rays are absorbed as they penetrate the weld. The weld itself absorbs most X-rays. If something less dense than the weld is present, such as a pore or lack of fusion defect, fewer X-rays are absorbed, darkening the film. If something more dense is present, such as heavy ripples on the weld surface, more X-rays will be absorbed, lightening the film.

Therefore, the foreign material's relative thickness and differences in X-ray absorption determine the radiograph image's final shape and shading. Skilled readers of radiographs can interpret the significance of the light and dark regions by their shape and shading. The X-ray image is a shadow of the flaw. The further the flaw is from the X-ray film, the fuzzier and larger the image appears. When thick material is x-rayed, flaws near the top surface may appear much larger than the same-sized flaw near the back surface. Those skilled at interpreting weld defects in radiographs must also be very knowledgeable about welding.

Ultrasonic Inspection (UT)

Ultrasonics is a fast and relatively low-cost nondestructive testing method that employs electronically produced, high-frequency sound waves (roughly 1/4 million to 25 million cycles per second) that penetrate metals and many other materials at speeds of several thousand feet (meters) per second. A portable ultrasonic inspection unit is shown in Figure 14-21.

The two types of ultrasonic equipment are pulse and resonance. The *pulse-echo* system, most often employed in the welding field, uses sound generated in short bursts or pulses. Since the high-frequency sound is at a relatively low power, it has little ability to travel through air, so it must be conducted from the probe into the part through a medium such as oil or water.

Sound is directed into the part with a probe held at a preselected angle or in a direction in which flaws will reflect some energy back to the probe. These ultrasonic devices operate very much like depth sounders or "fish finders." The speed of sound through a material is a known quantity. These devices measure the time required for a pulse to return from a reflective surface. Internal computers calculate the distance and present the information on a cathode ray tube, where an operator can interpret the results. The signals can be "monitored" electronically to operate alarms, print systems, or recording equipment. Sound not reflected by flaws continues into the part. If the angle is correct, the sound energy will be reflected to the probe from the opposite side. Flaw size is determined by plotting the length, height, width, and shape using trigonometric rules.

Leak Checking

Leak checking can be performed by filling the welded container with either a gas or liquid. Additional pressure may or may not be applied to the material in the weldment. Water is the most frequently used liquid although sometimes a liquid with a lower viscosity is used. If gas is used, it may be either a gas that can be detected with an instrument when it escapes through a flaw in the weld or an air leak that is checked with bubbles.

Figure 14-21 Portable ultrasonic inspection unit. (Courtesy of Magnaflux Corporation)

Hardness Testing

Hardness is the resistance of metal to penetration and is an index of the wear-resistance and strength of the metal. Hardness tests can determine the relative hardness of a weld with a base metal. The two types of hardness-testing machines in common use are the Rockwell and the Brinell testers. The Rockwell hardness tester uses a 120° diamond cone for hard metals and a 1/16″ (1.58 mm) hardened steel ball for softer metals (Figure 14-22). The method is based upon resistance-to-penetration measurement. The hardness is read directly from a dial on the tester. The depth of the impression is measured instead of the diameter. The tester has two scales—the B-scale and the C-scale—for reading hardness. The C-scale is used for harder metals, and the B-scale for softer ones.

The Brinell hardness tester measures the resistance of a material to the penetration of a steel ball under constant pressure (about 3,000 kilograms) for a minimum of approximately 30 seconds. The diameter is measured microscopically, and the Brinell number is checked on a standard chart. Brinell hardness numbers are obtained by dividing the applied load by the area of the surface indentation.

SUMMARY

Weld quality cannot be "tested into" anything, even if every part is tested. Quality must be *built* into the weldment because inspection only finds flaws and defects; it does not remove them. As you develop

Figure 14-22 Rockwell hardness tester. (Courtesy of Clark Instrument, Inc.)

your welding skills, you should work as if all of your welds will be subjected to an extensive testing program. Doing this will improve your skills and enable you to make quality welds when required.

REVIEW QUESTIONS

1. Where are noncritical welds found?
2. What determines the level and intensity of testing and inspection?
3. What are the two major classifications of testing products?
4. What type of testing is used as a baseline evaluation to establish welding procedure specifications?
5. What is a *discontinuity*?
6. What is the difference between a *discontinuity* and a *defect*?
7. What is *porosity*?
8. What causes nonmetallic inclusions?
9. When does inadequate joint penetration occur?
10. What is *interpass cold lap*?
11. When does *overlap* occur?
12. How can *crater cracks* be minimized?
13. What causes *lamination*?
14. What is the purpose of *fatigue testing*?
15. What are the two purposes for specimens to be tested by *etching*?
16. What is the *Izod test*?
17. What is *nondestructive testing*?
18. What is *magnetic particle inspection*?

APPENDIX

I. CONVERSION OF DECIMAL INCHES TO MILLIMETERS AND FRACTIONAL INCHES TO DECIMAL INCHES AND MILLIMETERS

Inches dec	mm	Inches dec	mm	Inches frac	dec	mm	Inches frac	dec	mm
0.01	0.2540	0.51	12.9540	1/64	0.015625	0.3969	33/64	0.515625	13.0969
0.02	0.5080	0.52	13.2080	1/32	0.031250	0.7938	17/32	0.531250	13.4938
0.03	0.7620	0.53	13.4620						
0.04	1.0160	0.54	13.7160	3/64	0.046875	1.1906	35/64	0.546875	13.8906
0.05	1.2700	0.55	13.9700	1/16	0.062500	1.5875	9/16	0.562500	14.2875
0.06	1.5240	0.56	14.2240						
0.07	1.7780	0.57	14.4780	5/64	0.078125	1.9844	37/64	0.578125	14.6844
0.08	2.0320	0.58	14.7320						
0.09	2.2860	0.59	14.9860	3/32	0.093750	2.3812	19/32	0.593750	15.0812
0.10	2.5400	0.60	15.2400	7/64	0.109375	2.7781	39/64	0.609375	15.4781
0.11	2.7940	0.61	15.4940						
0.12	3.0480	0.62	15.7480	1/8	0.125000	3.1750	5/8	0.625000	15.8750
0.13	3.3020	0.63	16.0020	9/64	0.140625	3.5719	41/64	0.640625	16.2719
0.14	3.5560	0.64	16.2560						
0.15	3.8100	0.65	16.5100	5/32	0.156250	3.9688	21/32	0.656250	16.6688
0.16	4.0640	0.66	16.7640						
0.17	4.3180	0.67	17.0180	11/64	0.171875	4.3656	43/64	0.671875	17.0656
0.18	4.5720	0.68	17.2720	3/16	0.187500	4.7625	11/16	0.687500	17.4625
0.19	4.8260	0.69	17.5260						
0.20	5.0800	0.70	17.7800	13/64	0.203125	5.1594	45/64	0.703125	17.8594
0.21	5.3340	0.71	18.0340	7/32	0.218750	5.5562	23/32	0.718750	18.2562
0.22	5.5880	0.72	18.2880						
0.23	5.8420	0.73	18.5420	15/64	0.234375	5.9531	47/64	0.734375	18.6531
0.24	6.0960	0.74	18.7960	1/4	0.250000	6.3500	3/4	0.750000	19.0500
0.25	6.3500	0.75	19.0500						
0.26	6.6040	0.76	19.3040	17/64	0.265625	6.7469	49/64	0.765625	19.4469
0.27	6.8580	0.77	19.5580	9/32	0.281250	7.1438	25/32	0.781250	19.8437
0.28	7.1120	0.78	19.8120						
0.29	7.3660	0.79	20.0660	19/64	0.296875	7.5406	51/64	0.796875	20.2406
0.30	7.6200	0.80	20.3200	5/16	0.312500	7.9375	13/16	0.812500	20.6375
0.31	7.8740	0.81	20.5740						
0.32	8.1280	0.82	20.8280	21/64	0.328125	8.3344	53/64	0.828125	21.0344
0.33	8.3820	0.83	21.0820	11/32	0.343750	8.7312	27/32	0.843750	21.4312
0.34	8.6360	0.84	21.3360						
0.35	8.8900	0.85	21.5900	23/64	0.359375	9.1281	55/64	0.859375	21.8281
0.36	9.1440	0.86	21.8440						
0.37	9.3980	0.87	22.0980	3/8	0.375000	9.5250	7/8	0.875000	22.2250
0.38	9.6520	0.88	22.3520	25/64	0.390625	9.9219	57/64	0.890625	22.6219
0.39	9.9060	0.89	22.6060						
0.40	10.1600	0.90	22.8600	13/32	0.406250	10.3188	29/32	0.906250	23.0188
0.41	10.4140	0.91	23.1140	27/64	0.421875	10.7156	59/64	0.921875	23.4156
0.42	10.6680	0.92	23.3680						
0.43	10.9220	0.93	23.6220	7/16	0.437500	11.1125	15/16	0.937500	23.8125
0.44	11.1760	0.94	23.8760						
0.45	11.4300	0.95	24.1300	29/64	0.453125	11.5094	61/64	0.953125	24.2094
0.46	11.6840	0.96	24.3840	15/32	0.468750	11.9062	31/32	0.968750	24.6062
0.47	11.9380	0.97	24.6380						
0.48	12.1920	0.98	24.8920	31/64	0.484375	12.3031	62/64	0.984375	25.0031
0.49	12.4460	0.99	25.1460	1/2	0.500000	12.7000	1	1.000000	25.4000
0.50	12.7000	1.00	25.4000						

For converting decimal-inches in "thousandths," move decimal point in both columns to left.

II. CONVERSION FACTORS: U.S. CUSTOMARY (STANDARD) UNITS AND METRIC UNITS (SI)

TEMPERATURE
Units

° F (each 1° change)	= 0.555° C (change)
° C (each 1° change)	= 1.8° F (change)
32° F (ice freezing)	= 0° Celsius
212° F (boiling water)	= 100° Celsius
−460° F (absolute zero)	= 0° Rankine
−273° C (absolute zero)	= 0° Kelvin

Conversions

° F to ° C _____ ° F − 32 = _____ × .555 = _____ ° C
° C to ° F _____ ° C × 1.8 = _____ + 32 = _____ ° F

LINEAR MEASUREMENT
Units

1 inch	= 25.4 millimeters
1 inch	= 2.54 centimeters
1 millimeter	= 0.0394 inch
1 centimeter	= 0.3937 inch
12 inches	= 1 foot
3 feet	= 1 yard
5280 feet	= 1 mile
10 millimeters	= 1 centimeter
10 centimeters	= 1 decimeter
10 decimeters	= 1 meter
1,000 meters	= 1 kilometer

Conversions

in. to mm _____ in. × 25.4 = _____ mm
in. to cm _____ in. × 2.54 = _____ cm
ft to mm _____ ft × 304.8 = _____ mm
ft to m _____ ft × 0.3048 = _____ m
mm to in. _____ mm × 0.0394 = _____ in.
cm to in. _____ cm × 0.3937 = _____ in.
mm to ft _____ mm × 0.00328 = _____ ft
m to ft _____ m × 32.8 = _____ ft

AREA MEASUREMENT
Units

1 sq in.	= 0.0069 sq ft
1 sq ft	= 144 sq in.
1 sq ft	= 0.111 sq yd
1 sq yd	= 9 sq ft
1 sq in.	= 645.16 sq mm
1 sq mm	= 0.00155 sq in.
1 sq cm	= 100 sq mm
1 sq m	= 1,000 sq cm

Conversions

sq in. to sq mm _____ sq in. × 645.16 = _____ sq mm
sq mm to sq in. _____ sq mm × 0.00155 = _____ sq in.

VOLUME MEASUREMENT
Units

1 cu in.	= 0.000578 cu ft
1 cu ft	= 1728 cu in.
1 cu ft	= 0.03704 cu yd
1 cu ft	= 28.32 L
1 cu ft	= 7.48 gal (U.S.)
1 gal (U.S.)	= 3.737 L
1 cu yd	= 27 cu ft
1 gal	= 0.1336 cu ft
1 cu in.	= 16.39 cu cm
1 L	= 1,000 cu cm

(continued)

II. CONVERSION FACTORS: U.S. CUSTOMARY (STANDARD) UNITS AND METRIC UNITS (SI) (continued)

1 L	= 61.02 cu in.
1 L	= 0.03531 cu ft
1 L	= 0.2642 gal (U.S.)
1 cu yd	= 0.769 cu m
1 cu m	= 1.3 cu yd

Conversions

cu in. to L _____ cu in. × 0.01638 = _____ L
L to cu in. _____ L × 61.02 = _____ cu in.
cu ft to L _____ cu ft × 28.32 = _____ L
L to cu ft _____ L × 0.03531 = _____ cu ft
L to gal _____ L × 0.2642 = _____ gal
gal to L _____ gal × 3.737 = _____ L

WEIGHT (MASS) MEASUREMENT

Units

1 oz	= 0.0625 lb
1 lb	= 16 oz
1 oz	= 28.35 g
1 g	= 0.03527 oz
1 lb	= 0.0005 ton
1 ton	= 2,000 lb
1 oz	= 0.283 kg
1 lb	= 0.4535 kg
1 kg	= 35.27 oz
1 kg	= 2.205 lb
1 kg	= 1,000 g

Conversions

lb to kg _____ lb × 0.4535 = _____ kg
kg to lb _____ kg × 2.205 = _____ lb
oz to g _____ oz × 0.03527 = _____ g
g to oz _____ g × 28.35 = _____ oz

PRESSURE and FORCE MEASUREMENTS

Units

1 psig	= 6.8948 kPa
1 kPa	= 0.145 psig
1 psig	= 0.000703 kg/sq mm
1 kg/sq mm	= 6894 psig
1 lb (force)	= 4.448 N
1 N (force)	= 0.2248 lb

Conversions

psig to kPa _____ psig × 6.8948 = _____ kPa
kPa to psig _____ kPa × 0.145 = _____ psig
lb to N _____ lb × 4.448 = _____ N
N to lb _____ N × 0.2248 = _____ psig

VELOCITY MEASUREMENTS

Units

1 in./sec	= 0.0833 ft/sec
1 ft/sec	= 12 in/sec
1 ft/min	= 720 in./sec
1 in./sec	= 0.4233 mm/sec
1 mm/sec	= 2.362 in./sec
1 cfm	= 0.4719 L/min
1 L/min	= 2.119 cfm

Conversions

ft/min to in./sec _____ ft/min × 720 = _____ in./sec
in./min to mm/sec _____ in./min × 0.4233 = _____ mm/sec
mm/sec to in./min _____ mm/sec × 2.362 = _____ in./min
cfm to L/min _____ cfm × 0.4719 = _____ L/min
L/min to cfm _____ L/min × 2.119 = _____ cfm

III. WELDING CODES AND SPECIFICATIONS

A *welding code* is a detailed listing of the rules or principles that are to be applied to a specific classification or type of product.

A *welding specification* is a detailed statement of the legal requirements for a specific classification or type of product. Products manufactured to code or specification requirements commonly must be inspected and tested to ensure compliance.

A number of agencies and organizations publish welding codes and specifications. The application of the particular code or specification to a weldment can be the result of one or more of the following requirements:

- Local, state, or federal government regulations
- Bonding or insuring company
- End user (customer) requirements
- Standard industrial practices

The three most popular codes are:

#1104, American Petroleum Institute—used for pipelines

Section IX, American Society of Mechanical Engineers—used for pressure vessels

D1.1, American Welding Society—used for bridges and buildings

The following organizations publish welding codes and/or specifications:

AASHT

American Association of State Highway and Transportation Officials
444 North Capitol Street, NW
Washington, DC 20001

AISC

American Institute of Steel Construction
1 East Wacker Drive
Chicago, IL 60601

ANSI

American National Standards Institute
11 W. 42nd Street
New York, NY 10036

API

American Petroleum Institute
2101 L Street, NW
Washington, DC 20005

AREA

American Railway Engineering Association
50 F. Street, NW
Washington, DC 20001

ASME

American Society of Mechanical Engineers
345 East 47th Street
New York, NY 10017

AWWA

American Water Works Association
6666 West Quincy Avenue
Denver, CO 80235

AWS

American Welding Society
550 NW LeJeune Road
Miami, FL 33126

AAR

Association of American Railroads
50 F. Street, NW
Washington, DC 20001

MIL

Department of Defense
Washington, DC 20301

SAE

Society of Automotive Engineers
400 Commonwealth Drive
Warrendale, PA 15096

IV. CHANGING WELDING VARIABLES FOR DESIRED RESULTS

WHEN TO TROUBLESHOOT	TUBE TO WORK DISTANCE	ELECTRODE WIRE-FEED SPEED	WIRE ELECTRODE VOLTAGE	WIRE ELECTRODE ANGLE	WIRE ELECTRODE TRAVEL SPEED	WIRE ELECTRODE DIAMETER	SHIELDING GAS
Poor Penetration	D	I/D	I/D	D	D	I/D/C	I/C
Rate of Deposit	I/D	I/D	NE	NE	I/D	I/D	NE
Porosity	I	D	D	I	D	NE	I
Electrode Stubbing	D	D	I	I	NE	NE	NE
Spatter Reduction	D	I	I	I	D	NE	C
Concave Face	I	I	I	I	I	D/C	C
Convex Face	D	D	D	D	D	I/C	C
Reduced Arc Blow	I	D	D	D	D	C	NE
Excessive Penetration	I	I/D	I/D	I	I	I/D/C	C
Weld Bead Size	I/D	NE	NE	C	I/D	NE	NE/LE
Weld Bead Width	I/D	NE	I/D	I/D	I/D	NE	I/D

KEY: / = Combination of Variables or the Change is Dependent on Other Preselected and/or Adjustable Variables
D = Decrease the Variable
LE = Little Effect on the Variable
C = Change the Variable
NE = No Effect on the Variable
I = Increase the Variable

V. GMAW TROUBLESHOOTING GUIDE

QUESTION	SOURCE	REMEDY
1. Arc Blow	A. DC magnetic field pulls on the arc. Arc wanders and is hard to control.	A. Copper or brass backing bars will help eliminate the arc "blow." B. Secure workpiece, check connections. C. Place workpiece connection close to the work. D. Change direction of the weld progression. E. Replace magnetized work bench.
2. Burn Through	A. Shielding gas B. Travel speed C. Current D. Root opening	A. Change from CO_2 to a mixture of Ar/CO_2 or Ar/O_2. B. Increase travel speed. C. Decrease current. D. Decrease root opening
3. Cracked Welds	A. Wrong electrode B. Poor joint design C. Bead too small, lack of reinforcement. D. Travel speed E. Poor welding technique F. Base metal	A. Check electrode selection chart for compatibility. B. Check root spacing and preparation of metal's edge. Preheat, postheat, maintain interpass temperature. Redesign weld joint. C. Increase wire-feed speed, change shielding gas, make weld bead contour more convex. D. Slow travel speed. E. Check electrode angles, maintain correct tube to work. F. Check quality of base metal.
4. Dirty Appearance	A. Dirty base metal B. Contaminated electrode wire C. Inadequate shielding gas cover	A. Clean base metal manually or with a chemical. B. Keep electrode package covered when on the wire-feed unit. Store electrode wire in the package container when not in use. Keep electrode packages in a dry location. C. Decrease tube to work. Increase shielding gas flow. Decrease the angle of the GMAW gun.
5. Excessive Spatter	A. Arc voltage	A. Check manufacturer's recommended voltage and make adjustments. Change shielding gas to AR/CO_2 or ArO_2
6. Excessively Wide Weld Bead	A. Arc length too long. B. Travel speed too fast. C. Welding current too high.	A. Decrease tube to work. B. Slow down travel speed. C. Decrease welding current.
7. Incomplete Weld Penetration	A. Poor joint design B. Travel speed too fast C. Tube to work too long. D. Poor welding technique.	A. Check root opening of joint. B. Slow down travel speed. C. Decrease tube to work distance. Increase the welding current. D. Adjust the GMAW gun angle.

(*continued*)

V. GMAW TROUBLESHOOTING GUIDE

QUESTION	SOURCE	REMEDY
8. Incomplete Fusion at the Weld Zone.	A. Tube to work too long. B. Base metal not cleaned. C. Poor joint preparation.	A. Decrease tube to work distance. B. Check joint, clean if needed. C. Review preparation methods, make corrections if necessary.
9. Lack of Fusion	A. Wrong polarity. B. Travel speed too slow. C. Welding current and voltage may be set too low. D. Welding technique may be too wide or too narrow. E. Base metal not clean.	A. Change to DCRP. B. Increase travel speed. C. Check manufacturer's recommendations, make adjustments. D. Correct technique. E. Properly prepare base metal.
10. Poor Welding Starts and Electrode Wire Stubbing	A. Voltage is too low or too high. B. Tube to work too long. C. Base metal needs cleaning.	A. Check welding variables, reset as needed. B. Decrease tube to work. C. Clean base metal.
11. Porosity throughout Weld Bead	A. Dirty or wet base metal. B. Dirty, rusty, or damp electrode wire. C. Poor shielding gas cover. D. Poor welder operator technique. E. Moisture	A. Properly clean base metal. B. Replace electrode wire. C. Clean nozzle. Decrease tube to work. Change electrode angle. Increase shielding gas flow. Check for obstructions. Check for shielding gas leaks. D. Correct angle of gun, run practice beads on scrap metal before welding on work. E. Use desiccant when storing electrode.
12. Porosity at start of weld bead	A. Ball of oxide not removed from end of electrode before the start of a weld. B. Moisture	A. Remove any balled up electrode metal from the end of the wire electrode. B. Replace electrode wire. Store electrode wire in a wire electrode rod oven. Use desiccant when storing electrode.
13. Porosity at end of weld bead	A. Rapid removal of the GMAW gun from the work area. B. Moisture	A. Hesitate for a moment after the weld is completed and the arc is extinguished before removing the gun. B. Use desiccant when storing electrode.
14. Undercutting	A. Welding current too high. B. Travel speed too fast. C. Poor gun manipulation and/or poor welder operator technique. D. Insufficient dwell time at the toes of the weld.	A. Check welding variables, make adjustments as needed. B. Slow down travel speed. C. Check transverse and longitudinal angles. D. Hesitate at the toe portion of the weld bead.

GLOSSARY

1F, plate Welding test position designation for a linear fillet weld; refers to a joint in which a weld is made in the flat welding position.

1G, plate Welding test position designation for a linear groove weld; refers to a joint in which a weld is made in the flat welding position.

2F, plate Welding test position designation for a linear fillet weld; refers to a joint in which a weld is made in the horizontal welding position.

2G, plate Welding test position designation for a linear groove weld; refers to a joint in which a weld is made in the horizontal welding position.

3F Welding test position designation used for a linear fillet weld; refers to a joint in which a weld is made in the vertical welding position.

3G Welding test position designation for a linear groove weld; refers to a joint in which a weld is made in the vertical welding position.

4F, plate Welding test position designation for a linear fillet weld; refers to a joint in which a weld is made in the overhead welding position.

4G Welding test position designation for a linear groove weld; refers to a joint in which a weld is made in the overhead welding portion.

100% root penetration See *root penetration.*

adjustable wrench Commercial tool that can be manually adjusted to accommodate a range of sizes.

all weld positions Joint orientation for a fillet weld or groove weld when testing. Includes flat, vertical, horizontal, and overhead welding positions.

alloy A substance with metallic properties and composed of two or more chemical elements of which at least one is a metal.

alloy steels Metal with one or more elements added, resulting in a significant change in the metal's properties.

alternating current (AC) Current that changes direction. Electrical current is the flow of electrons through a conductor. When alternating, the flow stops and changes direction, or reverses the flow, for each complete cycle. The U.S. standard is 60 cycles per second. Therefore, the current stops and starts 120 times per second. In theory electrons flow from negative to positive. As the current changes direction, the electrons do not but continue to flow from negative to positive.

American Iron and Steel Institute (AISI) Organization that classifies types of steel.

American National Standards Institute (ANSI) Voluntary membership organization that develops consensus standards nationally for a wide variety of devices and procedures.

American Petroleum Institute (API) Organization that issues codes and specifications for pipe welding and above-ground storage tanks.

American Society for Testing and Materials (ASTM) Voluntary membership organization concerned with materials and safety; a resource for sampling and testing methods, health and safety aspects of materials, safe performance guidelines, and effects of physical and biological agents and chemicals.

American Society of Mechanical Engineers (ASME) Organization that issues codes and specifications for pipe welding and boiler and pressure vessels.

American Welding Society (AWS) Organization that plays a major role in setting standards used throughout the welding industry.

amperage (A) Measurement of the total number of electrons flowing. Amperage controls the size of an arc.

antispatter Commercial product that prevents the buildup of weld spatter on or in the welding gun nozzle. Antispatter is available in a gel form, an aerosol spray, or a liquid dip.

arc burn See *arc strike*.

arc cutting A group of thermal cutting processes that severs or removes metal by melting it with the heat of an arc between an electrode and a workpiece.

arc gap See *arc length*. Arc gap is a nonstandard term for *arc length*.

arc length Distance from the tip or end of an electrode to the adjacent surface of the base metal's weld pool.

arc strike Discontinuity resulting from an arc, consisting of any localized remelted metal, heat-affected metal, or change in the surface profile of any metal object.

atmosphere Air surrounding the earth, measured in units of pressure equal to 14.69 pounds per square inch (psi). See *protective atmosphere*.

atmospheric contamination Removal of the protective envelope or atmosphere surrounding a weld zone provided by a shielding gas. Contamination of the weld zone results.

autodarkening welding lens Welding lens glass that automatically changes to the selected shade as quickly as 1/25,000th of a second.

automatic welding Welding with equipment that requires only occasional or no observation of the welding and no manual adjustment of the equipment controls.

auxiliary gas source Portable container of compressed shielding gas.

back-step welding Nonstandard term for a method of preventing hot spots in a weld. See *sequence*.

backgouging Removal of weld metal and base metal from the root side of a weld joint to facilitate complete fusion and complete joint penetration upon subsequent welding from that side.

backhand welding technique Welding technique in which a welding torch or gun is directed opposite to the progress of welding. The back of the welder's hand leads the direction of travel.

backing bar See *backing material*.

backing material A material or device placed against the back side of the weld joint or at both sides of a weld in electroslag and electrogas welding to support and retain molten weld metal. The material may be partially fused or remain unfused during welding and may be either metal or nonmetal.

backing strap See *backing material*.

backing strip See *backing material*.

barrel and pin Welding jargon that describes parts of the GMAW wire-feed unit. Holding device that keeps the reel, or coil, in place.

base metal The metal or alloy that is welded, brazed, soldered, or cut.

bead contour Nonstandard term that describes the shape of the face of a weld. See *concave, convex*.

bead pile-up Nonstandard term that describes excessive weld bead reinforcement.

bevel angle Angle between the bevel of a joint member and a plane perpendicular to the surface of the member.

beveled Edges of base metal prepared for welding by cutting them at an angle to accept a complete joint weld penetration.

bill of materials List of materials needed to construct a weldment or project.

bird's nest Nonstandard term that describes a malfunction of the wire-feed unit. The electrode wire tangles up around the drive and pressure rolls, giving the appearance of a bird's nest.

block sequence welding A combined longitudinal and cross-sectional sequence for a continuous multiple-pass weld in which separated increments are completely or partially welded before intervening increments are welded.

blow a fuse Nonstandard term describing the melting of the safety metallic stick inside the fuse housing due to overheating.

blow holes Nonstandard term that describes porosity. See *porosity*.

blueprint Instructions for fabricating a part or project.

bottom out The bottom of a range, the minimum descent or lowest level.

break line Ruled line with freehand zigzags showing long break in a part to conserve space on drawing.

bridge gaps See *bridging*.

bridging Nonapproved practice indicating that a space or opening exists that should be filled or closed (bridged) by welding. Usually not a close joint fit-up. Describes a weld joint clearance. A nonstandard term that describes a weld joint clearance that is to be filled by welding. See *gap*.

brittle Property of metal that cannot be deformed. A brittle material lacks plasticity.

building up Nonstandard term or welder's jargon sometimes used to refer to filling weld joints.

burn-back Nonstandard term that describes the wire electrode melting back toward the contact tube and preventing the wire from solidifying into the weld pool.

burn-off rate Nonstandard term for melting rate.

burn-through Nonstandard term that describes melt-through.

butt joint Joint between two members aligned approximately in the same plane.

cadence Rhythm or internal count used by welders to time welding travel speed and manipulation of an electrode.

carbon Nonmetallic element added to steel and cast iron that affects hardness and strength.

carbon dioxide (CO$_2$) Active shielding gas used in some arc welding processes.

carbon steel General term covering a large range of steels. See *low-carbon, medium-carbon,* and *high-carbon steel*.

cast Wire electrode forms a circle when a piece is dropped on a flat surface. The distance across the circle.

cascading (sequence) A combined longitudinal and cross-sectional sequence in which weld passes are made in overlapping layers.

cast iron Alloy of iron that contains carbon (up to 4.5%) plus silicon.

chromium Element added to steel to increase hardness and corrosion resistance. Chromium is the principal alloying ingredient of stainless steel.

circuit Path taken by electricity to complete a task. For example, when the GMAW gun is energized, an electrical circuit is formed by the wire electrode and the base metal, causing an electric arc and heat.

closed joint Nonstandard term that describes a joint with a root that is tight or closed together, preventing movement.

coalescence The growing together or growth into one body of the materials being welded. See *fusion welding*.

codes Welding requirements given on welding blueprints or engineering specifications to meet local, state, or federal safety standards.

coil 1 See *electrode packaging*.

cold lap Nonstandard term applied to incomplete fusion or overlap. Area of a weld that has not fused with the base metal. The area of a weld that has incomplete coalescence caused by insufficient application of heat to the base metal or poor welding technique.

cold lapping See *cold lap; overlap*.

cold start Nonstandard term for letting the welder know the start or beginning of the weld has not been fused to the base metal. See *cold lap*.

combination square Multisquaring tool used to determine 90° and 45° angles, centering of circles with a center-finding attachment, and degrees of a circle with a protractor attachment.

combustibles Any material able to catch fire and burn.

component Part of a whole.

composition Joining together of two or more substances.

compressed air Air held under pressure greater than atmospheric pressure. Air compressors supply air under pressure to operate air tools.

concave Term that describes the shape of the face on a weld. A depressed or inward-shaped surface.

conductor tube Component of the GMAW gun located between the body and the contact tube. Jargon for the conductor tube is *neck*.

conduit liner Internal component of the GMAW gun allowing the wire electrode to travel from the wire-feed rolls to the contact tube.

consistency Term that describes weld ripples that conform to one another. They hold to the same practice, repeatedly, in harmony.

constant-current power source (cc) An arc welding power source with a volt–ampere relationship, yielding a small welding current change from a large arc voltage change.

constant potential (CP) power source See *constant voltage (CV) power source*.

constant speed Refers to the constant-speed wire-feed unit. It is a trouble-free, solid-state, electronically controlled component of the GMAW process that provides the welder with regulated starts, automatically compensates for line-voltage fluctuations, and provides instantaneous response to wire electrode drag or restrictions. See *restriction*.

constant-voltage power source (CV) An arc welding power source with a volt–ampere relationship yielding a large welding current change from a small arc voltage change.

constituent Element or part of a metal alloy.

constituent of carbon steel A part of the composition of carbon steel. Alloys are parts of the makeup of some carbon steels.

contact tube Device that transfers current to a continuous electrode. A component of the wire-feed gun that energizes a wire electrode.

contaminated Indicates a dirty part, an impurity in the base metal or electrode, or the lack of a shielding gas or the shielding gas itself. Oil, paint, metal oxides, mill scale, or windy conditions that expel the gaseous shield from the weld zone or weld pool may cause this condition.

contraction Pulling together, becoming smaller, shrinking. The ability of a metal to decrease in all directions due to a decrease in temperature.

convex Term that describes the shape of the face on a fillet weld. A raised or outward contour surface.

copper cladding Covering or coating of copper, when referring to wire electrode.

corner joint Joint between two members located approximately at right angles to each other in the form of an *L*.

corrosion Decomposition or breaking down due to chemical reaction.

crack A fracture-type discontinuity characterized by a sharp tip and high ratio of length and width to opening displacement.

cracking Quickly opening and closing an auxiliary gas cylinder to clear the valve of foreign materials. Cracking the gas cylinder is jargon referring to removal of dirt or foreign material from the valve of a cylinder.

crater Depression in a weld face at the termination of a weld bead.

cubic feet per hour (cfh) Flow rate of gas through a regulator or flowmeter regulator.

current level Selected current for the desired metal thickness, electrode, electrode size, and shielding gas for a required weld.

cylinder (gas) Portable container used for the transportation and storage of a variety of compressed gases.

cylinder cracking A quick opening and closing of a cylinder to clear the valve of foreign materials.

cylinder protective caps Steel caps used on pressure cylinders to protect valves.

cylinder tension bolt Bolt that supplies tension on the reel of the wire-feed unit, producing or releasing drag to keep the reel from unwinding too easily.

cylinder wrench Special multiuse wrench for use on gas cylinders.

DCEP See direct current electrode positive.

defect Discontinuity that by nature or accumulated effort renders a part or product unable to meet minimum applicable acceptance standards or specifications. The term designates rejectability. See *discontinuity*.

defective See *defect*.

deoxidizer An element, such as silicon or phosphorus, added to filler metal to bond with oxygen, to keep oxygen from adversely affecting a weld.

deoxidizing See *deoxidizer*.

deposit Nonstandard term. See deposited metal. Refers to a weld metal being welded to a base metal by the addition of a filler metal to the weld pool. The weld bead is deposited into the base metal.

deposited metal Filler metal that has been added during welding, brazing, or soldering, or a surfacing metal that has been added during surfacing.

deposition rate Weight of a metal deposited in a unit of time.

depth of penetration Nonstandard term indicating depth that fusion extends into the base metal during welding.

desiccant Drying compound; absorbs moisture.

destructive testing (DT) Test of weld quality in which a joint is destroyed.

die Machine tool used to reduce the size of wire through the use of pressure and pulling forces.

direct current (DC) Flow of current in one direction, either from the work (base metal) to the electrode, or from the electrode to the work (base metal).

direct current electrode positive (DCEP) The arrangement of direct current arc welding leads in which the electrode is the positive pole and the workpiece is the negative pole of the welding arc.

discontinuity An interruption of the typical structure of a material, such as a lack of homogeneity in its mechanical, metallurgical, or physical characteristics. A discontinuity is not necessarily a defect.

distortion Refers to the uncontrolled movement of parts being welded due to the heat of welding. Distortion is also known as *warpage*. To prevent distortion when welding, fasten the welded part in a fixture.

drag restriction Excessive tension on a packaging reel, causing a wire electrode to require more force to unwind from the reel.

drive motor Electric motor found in the wire-feed mechanism used to propel the drive roller.

drive rolls Component of the wire-feed unit. Used to push or pull the wire electrode with the assistance of a pressure roller.

dross Scale or slag formation found on the back side of a kerf.

drum Type of filler metal package consisting of a continuous length of electrode wound or coiled in a cylindrical container (can). This can or container has a hole in the top for the wire electrode to exit to the wire-feed unit. The drum sits on the floor or a mounting stand and has a wire guide mechanism much like a large fishing pole reel to unravel the wire electrode from the container. The container should be handled in the upright position. This operation is used in production welding in which a large volume of electrode is required.

dual pulsed current Refers to the pulsed arc transfer mode in which one pulse of current is a spray transfer mode and the other pulse is at the higher end of the range of current.

ductility Ability of a metal to resist deformation when stretched, twisted, or flexed.

dwell time Length of time in the thermal spraying process that a surfacing material is exposed to the heat zone of a thermal spraying gun.

ear protection Any device that protects the ears from excessive noise; personal protective equipment (PPE).

edge joint Joint between the edges of two or more parallel or nearly parallel members.

egress route Escape route or path that workers use to exit from a work site.

electric shock Violent disturbance of the body caused by electric current passing through it.

electrode drag A restriction. An electrode is temporarily held back due to an interruption of the smooth flow of the electrode path on its way from the electrode package to the weld zone. See *restriction*.

electrode extension Length of wire electrode extending beyond the end of a contact tube.

electrode packaging Refers to the packaging or way in which a wire electrode is delivered and/or used, for example, wound on a spool, reel, or drum.

engine-driven motor generator power supply GMAW power source powered by a motor generator; used in field maintenance or portable welding operations.

entrance guide tube Opening or passageway set behind the drive rolls to guide a wire electrode into a drive roll mechanism. Guide tubes may be brass, copper, steel, plastic, or a short section of a liner.

equal leg Refers to equal distances on each leg from the joint root to the toe of a fillet weld.

expansion Movement of a base metal due to intense applied heat.

exit wire guide tube Opening or passageway set in front of drive rolls to control the exit from the rollers into a GMAW gun liner. Guide tubes may be brass, copper, steel, plastic, or a short section of the liner.

expansion Ability of a metal to grow in all directions due to an increase in temperature.

eye safety Pertains to general safety involving the eyes and area around the eyes. Can involve chemical, radiation, or projectile safety concerns.

eye sockets Area around the eye.

face reinforcement Weld reinforcement on the side of a joint from which welding was done.

face shield Protective device positioned in front of the eyes and over all or a portion of the face to protect the eyes and face.

ferrite Iron, Fe_3. Formation of pure iron in the grain structure of steel and cast iron.

ferrous metal Metal that contains iron.

filler metal Metal or metal alloy being added to a base metal when performing a weld.

fillet weld Weld of approximately triangular cross-section joining two surfaces at approximately right angles to each other in a lap joint, T-joint, or corner joint.

fillet weld gauge Tool used to measure the size of the legs of a fillet weld.

filter lens Nonstandard term for a filter plate; optical material that protects the eyes against excessive ultraviolet, infrared, and visible radiation. Also called *filter glass*.

fixture Device used to hold and maintain parts in proper relation to each other. Used as a clamping and squaring tool, usually custom made for a particular part or component.

flash protection Device that shields the eyes or exposed skin of the welder.

flat welding position The welding position used to weld from the upper side of a weld joint at a point where the weld axis is approximately horizontal and the weld face lies in an approximately horizontal plane. The workpiece is placed flat on a work surface for welding.

flaw An undesirable discontinuity. See *defect*; *discontinuity*.

flowmeter regulator Device used to monitor and control the flow rate of gas exiting the orifice of a portable cylinder or manifold. Used with a floating ball in a glass or plastic enclosure for monitoring purposes.

flush Refers to the alignment of metals or the face contour of a fillet weld after grinding.

flux Material that hinders or prevents the formation of oxides or other undesirable substances found in molten metals and on solid metal surfaces; flux dissolves or otherwise facilitates the removal of such substances. Generally found on the outside of SMAW coated electrodes or on the inside of a flux core wire electrode. Sometimes used in granulated form.

forehand technique Welding technique in which a welding torch or gun is directed toward the progress of welding. See also *push angle; work angle*.

free wheel Applies to a wire-feed unit electrode package being set too loosely, allowing the wire electrode to run off a reel.

fuse Safety device designed to break electrical connections by melting, thus shutting off current. Also describes the melding of metals as in "to fuse together."

fuse boxes Enclosure that houses fuses.

fusion welding The melting together of filler metal and base metal, or of base metal only, to produce a weld.

GMA See gas metal arc welding (GMAW).

GMAW-globular transfer See gas metal arc welding-globular transfer.

GMAW-P See gas metal arc welding - pulsed.

GMAW-S See gas metal arc welding - short circuit.

GMAW-spray transfer See *gas metal arc welding - spray transfer*.

GMAW welder Person who operates the equipment used in the gas metal arc welding process.

gap Nonstandard term to describe arc length, joint clearance, and root opening.

gas cylinder Portable container used for transportation and storage of compressed gas.

gas defuser Component of a GMAW gun that disburses shielding gas equally around a weld zone.

gas metal arc welding (GMAW) An arc welding process that uses an arc between a continuous filler metal electrode and a weld pool. The process is used with a shielding from an externally supplied gas and without the application of pressure.

gas metal arc welding - globular transfer Gas metal arc welding process variation in which molten metal transfers across the arc in large droplets. Used in combination with pulsed spray transfer.

gas metal arc welding - pulsed (GMAW-P) Gas metal arc welding process variation in which the current is pulsed.

gas metal arc welding - short circuit (GMAW-S) Gas metal arc welding process variation in which the consumable electrode is deposited during repeated short circuits.

gas metal arc welding-spray transfer Gas metal arc welding process variation in which molten metal transfers across a arc in small droplets.

globular transfer Transfer of molten metal in large drops from a consumable electrode across a arc.

goggles Protective glasses equipped with filter plates set in a frame that fits snugly against the face; used primarily with oxyfuel gas processes. Goggles can be clear or tinted.

gram Basic metric unit of measurement for mass and weight.

groove angle Total included angle of the groove between workpieces.

groove face Surface of a joint member included in the groove.

groove welds Weld made in a groove between the workpieces.

ground See *workpiece connection*. Nonstandard term for an electrical internal ground. Also refers to a surface finish made by grinding away metal.

gun angle Angle between the work and a welding torch (gun). This may be vertical, or a drag angle (backhand method), or a push angle (forehand method). A 10°–15° angle off the vertical is preferred for most applications to allow for the maximum amount of weld depth penetration.

gun assembly Refers to the external as well as the internal components of a GMAW torch.

gun body Plastic housing for a handhold; houses the trigger mechanism. Generally considered the main component of a gun assembly.

gun nozzle insulator Device in the GMAW gun that prevents the electrically live component from accidentally arcing by keeping the gun nozzle away from the contact tube.

ground clamp Nonstandard term for *workpiece connection*.

ground flush Refers to a contour finish of the face of a weld bead after a weld is completed. Surface metal of the weld bead area is removed by grinding to make it flush with the surface of the base metal.

ground lead Nonstandard and incorrect term for *workpiece lead*.

gun lead Cable from a wire-feed unit to a GMAW gun. This lead houses the electrode wire, liner, shielding gas, and electrical power cables.

gun liner Internal component of a GMAW gun lead. The liner is a flexible cable that allows a electrode wire to flow smoothly through. The liner may or may not need to be cut to fit.

gun neck, gun extension Component of a GMAW gun located between the gun body and nozzle. Permits spraying within confined areas or deep recesses. See *conductor tube*.

gun nozzle Device at the exit end of a torch or gun that directs shielding gas. Also referred to as a *gas nozzle*.

gun nozzle insulator Component of a GMAW gun that insulates the gun nozzle from the contact tube.

gun trigger A gun is operated by depressing the gun trigger, which is a component attached to a GMAW gun body. The gun trigger's microswitch or metal contact switch closes, completing a circuit and producing power to the gun.

handheld cutting torch Handheld, manually operated device, used in a number of cutting processes. See *oxyfuel cutting; plasma cutting*.

hard hats Personal protective equipment for the head; rigid construction.

hardness Resistance to penetration or denting.

hardness test Destructive test that measures the hardness (soundness) of a weld sample by indenting it.

hazardous Describes materials that are dangerous to the environment and people working in and around welding.

heat-affected zone Portion of a base metal whose mechanical properties or microstructure have been altered by the heat of welding, brazing, soldering, or thermal cutting.

helix Circular spiral or coil shape. Threads on a screw are helical. Slightly twisted as in a coiled spring or length of wire electrode.

high-carbon steel High-carbon steel has a carbon content of from 0.50% to 0.90%. It is hard and strong. Pre and post heating is required for most welding to prevent weld cracking.

horizontal welding position Welding position in which a weld face lies in an approximately vertical plane and the weld axis at the point of welding is approximately horizontal. The workpiece is positioned perpendicular to the work surface for welding.

hot wires Component of GMAW gun that makes it useable without a switch. When an electrode wire contacts the base metal, the wire begins to flow, establishing an arc at the moment of contact.

hydrogen entrapment Presence of hydrogen trapped as the weld pool solidifies. Pockets of hydrogen cause stresses that lead to small cracks in the weld metal that may later enlarge.

impact resistance Resistance to sudden force without fracture.

impurities Foreign substance on or entrapped in a weld or base metal. See *inclusions*.

inches per minute (ipm) Wire-feed speed, that is, how fast an electrode wire is consumed into a weld pool. Also used to measure how fast a wire electrode exits the contact tube of a GMAW gun.

inclusions Entrapped foreign solid material, such as slag, flux, tungsten, or oxide.

incomplete fusion Weld discontinuity in which fusion did not occur between weld metal and fusion faces or adjoining weld beads. See *cold lap; overlap*.

inert gas Gas that does not normally combine chemically with materials. Inert gases do not break down when ionized by an electric arc. Argon and helium are typical inert gases used as a shielding gas in welding.

infrared light (IR) Light that emits intense heat and causes painful burns to exposed skin.

inside corner joint Refers to a corner joint where a weld is performed on the inside of a joint.

integrity Character or construction of a material.

intermediate travel speed Preferred rate of travel for maximum weld penetration; speed that allows a welder to stay within the selected welding variables and control fluidity of a weld pool.

internal ground Electrical connection made inside a welding power source at a factory by a qualified technician.

International System of Units (SI) System of common units of measurement used worldwide. See SI Basic Units.

inverter power source Small, powerful welding power source.

iron Iron is one of the purist forms of the element iron used. It has a carbon content of from 0.0% to 0.003%. It is soft, easily formed, and will not harden.

iron carbon alloy A carbon steel that contains minimum amounts of alloys added to achieve desired effects.

ironworker Machine that combines many functions of tools such as a metal shear, punch press, angle shear, bar shear, and shearing of round stock. An ironworker can also be a person who works with iron and steel.

joint penetration Distance a weld metal extends from a weld face into a joint, exclusive of weld reinforcement.

joint configurations Various combinations of joint construction.

joint root That portion of a joint to be welded where the members approach closest to each other. In cross-section, a joint root may be either a point, a line, or an area.

kerf The width of a cut produced during a cutting process. The area where metal has been removed.

keyhole See *keyhole welding*.

keyhole welding A welding technique in which a concentrated heat source penetrates partially or completely through a workpiece, forming a hole (keyhole) at the leading edge of a weld pool. As the heat source progresses, the molten metal fills in behind the hole to form a weld bead.

kilogram Standard metric unit of measurement; measures weight in 1,000 grams.

lap joint A joint between two overlapping members in parallel planes.

layout Layout is the process of drawing lines or making marks on the material or parts to be used for a weldment from drawings, sketchs or other sources.

leathers Personal protective equipment; leather clothing worn by welders for protection from hot metal and sparks.

liter (litre) Basic metric unit of measurement for capacity or volume.

longitudinal angle Angle of a GMAW gun less than 90° between a perpendicular line to the workpiece; position of a GMAW gun.

low-carbon steel Low-carbon steel has a carbon content of from 0.003% to 0.30%. It is strong, formable and easily welded. Some times referred to as *mild steel*.

luminous intensity Basic metric unit of measurement; measures light.

MSDS See *material safety data sheets*.

magnifier lens Welding lens that enlarges viewing area.

manganese Grayish, metallic chemical element, used as an alloy to add strength and improve hardness and to remove impurities.

manipulate Motion made when welding with a handheld gun or GMAW gun. See *oscillate*.

mass Basic metric unit of measurement; measures heaviness.

material safety data sheets (MSDS) Information provided with welding equipment and supplies describing product composition and possible hazards.

mechanical property Ability of a metal to react to impact, twisting, pulling, indentation, bending, and other mechanical forces.

medium-carbon steel Medium-carbon steel has a carbon content of from 0.30% to 0.50%. It has high strength and is tough. To prevent post weld cracking, a controlled cooling rate is required for most welding.

melt-through Visible root reinforcement produced in a joint welded from one side.

melting point Temperature at which a wire electrode or base metal liquifies.

member AWS term for base metal, parent metal, plate, workpiece, or sections to be joined.

metal scribe Handheld tool used to scratch a mark on metal, making a visible line when laying out parts to be welded.

metal tongs Plier-type tool used to handle hot metal.

metallurgical change Change in the molecular structure of a metal due to heat treatment or additional alloys.

metallurgical characteristics Properties that a metal member offers; may be physical, mechanical, or chemical.

meter Standard metric unit of measurement for length.

Mig Nonstandard term used to describe the GMAW process. See *gas metal arc welding*.

mild steel See *low-carbon steel*.

mill scale Oxides that form on the surface of mild steel. A gray, blue, or black flake substance, very thin and easily removed. Mill scale can affect weld quality and must be removed.

mode Refers to a method or mechanism used to transfer metal; may be short circuit, spray, globular, pulsed, or the type of operation the process is used in, semiautomatic or automatic.

mode of metal transfer Mechanism used to transfer metal from an electrode to the work. See globular, pulsed arc, short circuit, and spray transfer. (See *AWS Welding Handbook*, Vol. 2, p. 112.)

multipass padding plate Series of welding beads layered on top of one another on a base metal plate.

multiprocessors Inverter or multiprocessor power source that can perform a variety of welding processes.

needlenose pliers Hand tool used to clean a GMAW gun nozzle of spatter and snip off the wire electrode.

needle scaler Air or electric tool used to remove slag, scale, or dross from a weld zone either before or after welding.

negative lead cable Welding lead cable for negative current that is attached to a base metal.

new metal Jargon that refers to metal in a weld zone that has not been touched by an arc but needs to be.

nibblers Handheld air or electric tool used for removing metal in small snips or bites. (Portable shear for light-gauge metals.)

nickel Alloying element that increases the hardness of steel. Nickel is often used in combination with chromium.

nod Head motion used by welders to lower a welding helmet without using the hands.

noise Unwanted sound, form of vibration transmitted through gases.

nondestructive examination (NDE) Act of determining the suitability of a material or component for its intended purpose using techniques that do not affect its serviceability.

nonferrous metals Metals containing no iron.

nozzle Component of GMAW gun housing contact tip. In arc spraying, the nozzle directs and forms the flow shape of atomized spray particles and the accompanying air or other gases.

OSHA See *Occupational Safety and Health Administration*.

Occupational Safety and Health Administration (OSHA) An agency in the U.S. Department of Labor with safety and health regulatory and enforcement authority for most U.S. industries and businesses.

open-butt joint Nonstandard term that describes a butt joint with an open root.

oscillate Motion a welder uses when using a handheld gun to produce a weld bead. See *manipulate*.

oscillation movement See *oscillate*.

other side Opposite side of a base metal plate where welding is taking place.

out-of-position welding Jargon for welding in other than the flat position.

outside corner joint Joint with two members located approximately at right angles. The joint is welded on the exposed or exterior side (outside).

overlap A nonstandard term when used for incomplete fusion. In fusion welding, overlap is the protrusion of weld metal beyond the weld toe or weld root.

oxidation Formation of oxides that can affect weld quality. Breaking down or decomposing process of steel. Rusting is an example of slow oxidation. Burning is an example of rapid oxidation.

oxide Compound of oxygen and iron that forms on steel and can affect the strength of a weld. Oxides can cause weld failure if not removed.

oxygen cutting (OC) Group of thermal cutting processes that sever or remove metal by means of a chemical reaction between oxygen and a base metal at elevated temperatures. The necessary temperature is maintained by the heat from an arc, an oxyfuel gas flame, or other source.

oxyfuel cutting torch A device used for directing the preheating flame, produced by the controlled combustion of oxygen and a fuel gas mixture, and to direct and control the cutting oxygen stream.

PPE Personal protective equipment used to protect a welder from physical harm when welding.

packaging Material with which, or in which, a wire electrode is contained.

parameters Boundaries or settings. Set to work within parameters may vary depending on the welding process and the size and type of electrode.

peak out Refers to electrical controls, the top of a range, the maximum climb, or highest level.

periodic Appearing or recurring at regular intervals

periodic maintenance Scheduled maintenance performed to ensure troublefree operation of GMAW equipment.

personal protective equipment (PPE) Various pieces of equipment or apparel that protect a welder from injury while welding.

phosphorus Nonmetallic element added as an alloy to increase the strength and corrosion resistance of some steels.

physical properties A material's characteristics, measured against mechanical force.

pinch effect High or increased current flow, combining with a high percentage of argon shielding gas; causes a squeezing of the molten ball at the end of an electrode wire during the spray transfer mode.

plasma arc cutting torch A device used to create a high velocity jet of ionized gas (plasma) and direct that plasma column through a constricting orifice so that it is concentrated enough to cut through metal.

plastic state Refers to molten metal just before it solidifies—not liquid, but not solid; soft and pliable.

pneumatic Powered by air pressure; generally refers to tools or equipment with air being the driving force of a tool.

porosity Cavity-type discontinuities formed by gas entrapment during solidification or in a thermal spray deposit.

positive lead cable Welding lead cable for positive current that is attached to the positive or plus (+) terminal of the power supply of a GMAW wire-feed unit and gun.

power cord Cable leading from a power supply to an electrical receptacle located at the source of electrical power.

power lead Cable connecting a torch to electric current in gas metal arc welding.

power source Apparatus for supplying current and voltage suitable for welding. Apparatus responsible for providing the amperage and voltage needed for a welding operation; the main component of a GMAW package. Also called by the nonstandard terms *welding machine, welding unit,* or *welder.*

power supply Nonstandard term for a welding machine or welder. *Power source* is the correct term.

preheat The heat applied to a base metal or substrate to attain and maintain preheat temperature.

preselected variables Adjustments made to the welding equipment before welding commences.

pressure regulator See *flowmeter regulator.* Pressure sensing device that regulates the flow of shielding gas without a floating ball but uses a dial and needle as an indicator showing cylinder pressure and flow pressure.

primary adjustable variables Variables that control the process after preselected variables have been set. They control the weld bead, the arc voltage, travel speed, and welding current. Some variables can be adjusted by the welder as the weld progresses.

primary voltage High voltage—230-460 volts.

protective atmosphere Gas or vacuum envelope surrounding the workpieces, used to prevent or reduce the formation of oxides and other detrimental surface substances and to facilitate their removal.

protective envelope Cloud of shielding gas that surrounds an arc when welding, keeping contaminated atmosphere from entering a weld zone.

protective garments Personal protective equipment that protect a welder from heat, sparks, and light radiation. See *leathers* or *personal protective equipment.*

pulsed arc transfer (GMAW-P) Gas metal arc welding process variation in which the current is pulsed.

qualified repair technician An individual who has been trained to service and repair a welding power source and related equipment.

quality control Inspection, testing, and qualification in manufacturing and maintenance processes.

quality welds Welds that meet an accepted standard of excellence.

quench The act of cooling metal with a variety of media.

quick-change welding lens Liquid-crystal filter lens capable of changing from a shade no. 3 to a preselected shade. As an arc is struck, the filter lens darkens in less than 1/25,000 of a second. See *auto darkening welding lens.*

radiation Transmission of energy by electromagnetic waves longer than visible light.

rays See *waves.* Nonstandard term for radiant energy of any wavelength.

recommended weld position Position in which a workpiece is placed for welding.

reel A type of filler metal package consisting of a continuous length of electrode wound on a cylinder, called a *barrel.* This reel is flanged at both ends and extends past the inside diameter of the barrel. It has a spindle hole and mounts on or inside the wire-feed unit to supply the continuous-wire electrode. Sometimes called the *spool.* See *drum; spool.*

reel retainer Refers to a slip pin or related device that prevents a wire reel (barrel) from falling off of the wire-feed unit.

restriction Any drag or pinch found in the wire-feed system of a wire feeder, either in the drive roll mechanism, the gun assembly, or the contact tip. A restriction will inhibit the wire electrode from sliding smoothly through the drive rolls, conduit liner, and contact tip, preventing or restricting the weld from being performed correctly.

reverse polarity Nonstandard term for *direct current electrode positive*. DCRP, DCEP.

ripples Nonstandard term referring to the shape of a deposited weld bead during the weld process or after solidification has taken place.

robot Programmable tool designed to preform many types of operations that are repeated by a tool (machine) without interaction of a machine operator. A programmed sequence of events allows the function to be preformed exactly. Welding may be one of the functions of a robot.

root Nonstandard term when used for *joint root* and *weld root*.

root bead Weld bead that extends into or includes part or all of a joint root.

root face That portion of a groove face within a joint root.

root opening Separation at the joint root between workpieces.

root penetration, 100% Distance a weld metal extends into a joint. Complete penetration means 100% penetration.

run-off Refers to a wire electrode free wheeling and coming off or out of a weld package while on a wire-feed unit. See *free wheel*.

rupture In destructive weld testing, bending a weld sample until the welded joint bends, breaks, or opens up.

SI Basic Units International basic metric units, electrical current, length, luminous intensity, mass, substance, temperature, time.

SI Prefixes Prefixes used with basic metric units of measurement: kilo - 1,000 (one thousand); hecto - 100 (one hundred); deca - 10 (ten); deci - 1/10 (one tenth); centi - 1/100 (one one-hundredth); milli - 1/1000 (one one-thousandth).

scratch started Refers to GMAW guns that have hot wires that must be scratch started to start the flow of electrode wire. As soon as wire electrode contacts the base metal, the arc is visible, and the wire begins to flow.

second Standard unit of measurement; measures time.

secondary voltage Occurs when you interrupt the electrical circuit established during welding; touching both the work and an uninsulated part of the equipment at the same time.

secure Fasten metal parts in position to prepare them for welding. See *fixture*.

selective positioning of weld beads Positioning weld beads to prevent hot spots from developing during welding. See *back-step welding; sequence*.

self-testing Testing oneself on abilities acquired during a welding training program.

semiautomatic welding process Manual welding with equipment that automatically controls one or more welding conditions. A welder manually manipulates a welding gun to achieve a desired weld, while the continuously fed wire electrode is automatically added to a weld pool.

sequence Method of reducing concentration of heat or hot spots in one area of a weld zone. See *back-step welding; selective positioning of weld beads*.

shades Various levels of dark or light filter lenses used during the welding process.

shielding gas Protective gas or envelope surrounding a weld pool. This protective gas prevents a weld zone from atmospheric contamination.

short-circuit transfer Metal transfer in which molten metal from a consumable electrode is deposited during repeated short circuits 20–200 times per second. See *gas metal arc welding*.

side-cutter attachment A cutting blade found on a needle-nose plier. Used to cut wire electrode.

silicone Non-metallic element that can increase the hardness of steel, improve its mechanical properties, and reduce oxidation.

single-bevel butt joint Nonstandard term for *single-bevel groove*.

single-bevel groove Type of groove weld in which a single end of one plate is grooved and the other plate is square.

single V-groove Groove weld that is made from one side with both plates beveled to form a V.

skip welding Positioning weld beads to prevent hot spots from developing during welding. See *sequence*.

slag Nonmetallic product resulting from the mutual dissolution of flux and nonmetallic impurities in some welding and brazing processes.

soapstone A marking device much like a crayon that is used to lay out metal for welding, cutting, drilling, or machining.

solid wire Term that describes a wire electrode. Also known as *hard wire*.

sound weld Weld that has joined together base metal and filler metal in a manner that is acceptable in appearance and strength. See *quality welds*.

spatter Metal particles expelled during fusion welding that do not form a part of the weld.

specifications Requirements that may work together with codes. Specifications are generally determined by fabricators or design engineers.

spool Type of filler metal package consisting of a continuous length of electrode wound on it, then placed on a cylinder called a *barrel*. This spool is flanged at both ends and extends past the inside diameter of the barrel. It has a spindle hole and mounts on or inside the wire-feed unit to supply a continuous wire electrode. Sometimes called a *reel*. A wire drum is used for production welding. See *drum; reel*.

spray transfer Metal transfer in which molten metal from a consumable electrode is propelled axially across an arc in small droplets. See *gas metal arc welding*.

square Fasten metal parts in position so that the parts form the required angles to prepare them for welding. See *combination square; fixture*.

square groove See *square groove weld joint*.

square groove weld joint Type of groove weld in which both plates are or remain square.

staggering Intermittent welds on both sides of a joint in which the weld increments on one side are alternated with respect to those on the other side. Positioning weld beads to prevent hot spots during welding.

stand-off distance The distance from a nozzle to a workpiece.

start Point at which an arc is struck, and a weld bead begins to form.

sterile eye wash solution Commercial liquid for flushing the eyes during an accident.

stick out Distance a wire electrode extends from the end of the contact tube to the base metal (work), including the actual arc length. Also known as *tube to work distance*. In welder's jargon the nonstandard term is *tip to work*.

straight edge Measuring device used to visually check trueness of an edge in question with a known true straight edge. The longer the straight edge, the more accurate the measurement.

strain Deformation of a material due to stress. Strain can lead to weld failure.

strength Ability to resist strain or deformation. Refers to the ability of a material to resist loads.

stress Internal or external forces acting on a material. Stress can result in strain or deformation.

stringer weld bead Weld bead deposited with little side-to-side movement (without weaving).

substance Basic metric unit of measurement; measures matter.

sulfur Element sometimes found in small quantities mixed with manganese to improve machining.

surface reinforcement The allowable weld metal on a surface, either root or face, in excess of the required amount needed to fill a joint. More than the allowable by code or specification would be considered excessive and is not acceptable.

synchronize Happening at the same rate; in harmony.

tack weld Weld that holds the parts of a weldment in proper alignment until the final welds are made.

tagged Refers to idenifying equipment that needs repair. A tag describing the malfunction or problem is attached to a machine.

tee joint Joint between two members located approximately at right angles to each other in the form of the capital letter *T*.

temperature gradient Drop in temperature from a weld zone away from a weld pool.

tempering Heat treatment by which hardened steel is toughened.

tensile strength Resistance to being pulled apart.

tension adjustment Refers to the tension applied to a component of a wire-feed unit to produce a drag or to release drag on the wire electrode.

thermal conductivity A measure of the rate at which a metal transfers heat.

thermal cutting Group of cutting processes that severs or removes metal by localized melting, burning, or evaporating of the workpieces. See *arc cutting; oxygen cutting.*

tie-in Nonstandard term that describes the fusing of weld beads, side by side as in a weld pad, or the start of a weld bead. Using the crater of an existing bead to connect beads in a straight line or to tie them to one another.

time Basic unit of measurement; measures duration or span.

toes Junction of a weld face and a base metal.

tongs Blacksmith tool used to hold hot metal. Somewhat like a giant pair of pliers.

torch In gas metal arc welding, torch refers to the gun. Other welding processes have different definitions of a torch. See *gun assembly; gun body.*

toughness Resistance to fracture from a constant force.

trail behind an arc Refers to the flow of a weld pool in which a welder maintains contact with the wire electrode with both the plates of the workpiece at the same time and causes the weld pool to trail behind the arc to form a sound weld.

transformer–rectifier power source Preferred power source for use in a shop or factory where electric power is accessible.

transverse angle Approximately one-half the included angle between the plates that form a joint.

travel angle Angle of a gun in relation to the direction of travel along a joint. This angle can also be used to partially define the position of guns, torches, rods, and beams.

travel speed Speed of weld progression along a weld seam. Generally determined in inches per minute (ipm).

triggering Short bursts of a welding arc. An on-and-off action controlled by a welder. Used on thin materials or large openings.

troubleshooting Looking for potential corrective measures when a welding problem is detected.

tube to work Distance from a contact tube to a workpiece.

tungsten Chemical element with the highest melting point of any known metal—6, 170° F—and a boiling point temperature of 10,700° F.

ultraviolet light (UV) Light ray beyond the visible spectrum. Causes burns to the skin without immediate pain and, for this reason, is considered the most dangerous of all forms of light.

undercut Groove melted into a base metal adjacent to a weld toe or weld root and left unfilled by weld metal, leaving a rough, jagged edge.

underfill Condition in which a weld face or root surface extends below an adjacent surface of a base metal but not undercut. A smooth edge is left.

unequal leg Refers to the leg size of a fillet weld. One leg may be longer or shorter than the other.

unigoggles Type of face shield in which goggles and face shield form a single unit.

user standards Requirements set by a fabricator or end user of a product.

vanadium A metallic element used as an alloy to increase hardness.

very high carbon steel Steel that contains 0.65% (0.0065) to 1.5% (0.015) carbon.

visible light Light you can see; falls midway in the electric–magnetic spectrum chart.

volatile Capable of exploding.

voltage (V) Measurement of electrical pressure. Voltage controls the maximum gap the electrons can jump to form an arc.

warpage Refers to the movement of parts being welded. Warpage is also known as *distortion.* To prevent warpage when welding, the welded part should be fastened in a fixture. Preheating and postheating techniques may be needed.

wattage (W) Watts are calculated by multiplying volts times amperes. Wattage is the measurement of the amount of power in an arc. The wattage (power) being put into a weld per inch controls the width and depth of the weld bead.

waves Refers to light radiation, infrared, ultraviolet, or visible light.

weave bead Type of weld bead made with transverse oscillation. Any one of several patterns produced with pronounced side-to-side movement.

weaving motion Refers to a type of weld bead made with transverse oscillation.

weld A localized coalescence of metals or non-metals produced by heating the material to the weld temperature, with or without the application of pressure and with or without the use of a filler metal. See *coalescence.*

weld bead Weld from a single pass. See *stringer weld bead; weave bead.*

weld characteristics Qualities comprising a weld. See *sound weld; quality welds.*

weld contour Shape of a weld reinforcement. See *concave; convex.*

weld face Exposed surface of a weld on the side on which welding was done. Jargon for *weld face* is *crown.*

weld flash Welding light that is reflected or inadvertently seen from a distance.

weld groove Channel in the surface of a base metal between two joint members.

weld interface Interface between weld metal and base metal in a fusion weld. See *tie in.* Interface refers to the fusing of metal. See *coalescence; fusion.*

weld joints Junction of members or the edges of members that are to be joined or have already been joined.

weld legs A fillet weld leg is the distance from a joint root to the toe of a fillet weld.

weld pass Single progression of welding along a joint. The result of a pass is a weld bead or layer. More than one pass is a multipass weld.

weld pool Localized volume of molten metal in a weld prior to its solidification as weld metal. A nonstandard term for *pool* is *weld puddle.*

weld root Point at which a weld penetrates into a joint root; the farthest point from a weld face.

weld sample Specimen or coupon. Refers to a section of a weld being tested.

weld size Refers to the size of a weld bead deposited, leg size, depth of penetration, and amount of face reinforcement.

weld spatter Expelled metal particles originating from a weld zone that are not part of a weld and do not form part of a weld.

weld symbols Graphical characters connected to the welding symbol indicating the type of weld.

weld throat Shortest distance between a weld root and a weld face.

weld toes Junctions of a weld face and a base metal.

weld zone Nonstandard term for surface area immediately surrounding a weld pool.

weldability The capacity of material to be welded under imposed fabrication conditions into a specific, suitably designed structure and to perform satisfactorily in the intended service.

welder Machine operator; one who performs the GMAW process. Also a nonstandard term sometimes used for a power supply.

welder certification Written certification that a welder has produced welds meeting a prescribed standard of welder performance.

welder's cap Personal protective head covering worn by a welder to prevent sparks and hot metal from reaching the hair, neck, and ears. Also makes an operator more visible to other workers.

welder's hood Nonstandard term for *welding helmet.*

welder's jargon Nonstandard welding terms used by welders.

welder's leathers Protective clothing made from leather or other fire-resistant fabrics. See *personal protective equipment; leathers.*

welding Localized coalescence of metals or non-metals produced either by heating the materials to the welding temperature, with or without the application of pressure, or by the application of pressure alone and without the use of filler material. Also, controlled short circuit by which an electric current causes intense heat in a weld pool between an electrode and a base metal.

welding arc Controlled electrical discharge between an electrode and a workpiece that is formed and sustained by the establishment of a gaseous conduction medium called an arc plasma.

welding gun Device used in GMAW to direct a wire electrode at a base metal and to transfer current to the arc, direct the shielding gas, and to manipulate the weld pool to perform the desired weld. Also called a *welding torch.*

welding helmet Device equipped with a filter plate designed to be worn on the head to protect the eyes, face, and neck from arc radiation, radiated heat, spatter, or other harmful matter expelled during welding and cutting processes.

welding leads Cables connecting an electrode GMAW gun and a ground clamp to a power source. They are constructed of fine copper wires stranded together and enclosed in a flexible synthetic rubber or plastic jacket to provide toughness, flexibility, electrical, and heat-resistant qualities.

welding position The relationship between a weld pool, joint, joint member, and welding heat source during welding.

welding power source Apparatus for supplying current and voltage suitable for welding.

welding technique Details of a welding procedure that are controlled by a welding machine or welder.

welding terminals Positive and negative connections located on a power source to which the GMAW wire-feed unit and base metal are connected with welding leads. See *welding leads*.

welding torch Device used in GMAW to direct a wire electrode at a base metal and to transfer current to the arc, direct the shielding gas, and manipulate the weld pool to perform the desired weld. Also called a *welding gun*.

welding variables Adjustments to a power source and related equipment; also electrode manipulation.

weldment Assembly whose component parts are joined by welding.

weldor Refers to a welder (nonstandard term).

welper Specialty tool used to perform maintenance functions on a GMAW gun.

wetting ability Ability of a wire electrode to liquefy and to pool, spreading itself and adhering to a base metal, forming a coalescence.

whiskers Short stubs of wire electrode that protrude through the root side of a weld joint after welding. Undesirable; shows a need for improved welder technique.

wire electrode Form of welding filler metal, normally packaged as coils, spools, reels, or drums. Known as *continuous wire electrode*.

wire-feed gun assembly Welding gun.

wire-feed speed Rate at which wire is consumed in an arc when welding.

wire-feed unit Component of the GMAW process. Provides a steady supply of wire electrode to a weld pool.

wire guide Component of a wire-feed unit. Device that helps steer the wire electrode into and out of the pressure and drive rolls.

wire melting rate Rate at which an electrode wire is consumed into a weld pool; measured in inches per minute (ipm).

work Material being welded; also known as *parent metal* or *base metal*. See *workpiece*.

work angle An angle less than 90° between a line perpendicular to a major workpiece surface and a plane determined by the electrode axis and the weld axis. In a tee joint or corner joint, the line is perpendicular to the nonbutting member. This angle can also be used to partially define the position of guns, torches, rods, and beams.

workpiece Part that is welded, brazed, soldered, thermal cut, or thermal sprayed.

workpiece connection Connection of a workpiece lead to a workpiece.

zinc Silver-white metallic element used as a protective coating for galvanizing metal. When combined with copper, it forms brass.

INDEX